Metaheuristics and Reinforcement Techniques for Smart Sensor Applications

This book discusses the fundamentals of wireless sensor networks, and the prevailing methods and trends of smart sensor applications. It presents analytical modelling to foster the understanding of network challenges in developing protocols for next-generation communication standards.

- Presents an overview of the low-power sensor, network standards, design challenges, and sensor network simulation
- Focuses on clustering methods available for wireless sensor networks to tackle energy hole problems, load balancing, and network lifetime enhancements
- Discusses enhanced versions of energy models enriched with energy harvesting
- Provides an insight into coverage and connectivity issues with genetic metaheuristics, evolutionary models, and reinforcement methodologies designed for wireless sensor networks
- Includes a wide range of sensor network applications and their integration with social networks and neural computing

The reference book is for researchers and scholars interested in Smart Sensor applications.

Metaheuristics and Reinforcement Techniques for Smart Sensor Applications

Adwitiya Sinha, Manju,
and Samayveer Singh

CRC Press
Taylor & Francis Group
Boca Raton London New York

CRC Press is an imprint of the
Taylor & Francis Group, an **informa** business

A CHAPMAN & HALL BOOK

Designed cover image: ShutterStock

First edition published 2025
by CRC Press
2385 NW Executive Center Drive, Suite 320, Boca Raton FL 33431

and by CRC Press
4 Park Square, Milton Park, Abingdon, Oxon, OX14 4RN

CRC Press is an imprint of Taylor & Francis Group, LLC

ISBN: 9781032542355 (hbk)
ISBN: 9781032548524 (pbk)
ISBN: 9781003427780 (ebk)

DOI: 10.1201/9781003427780

Typeset in Times
by Deanta Global Publishing Services, Chennai, India

Contents

CHAPTER 1

CHAPTER 2

CHAPTER 3

CHAPTER 4

CHAPTER 5

CHAPTER 6

CHAPTER 7

Preface

Wireless sensor networks have become a vibrant technology that encompass a wide variety of potential applications in civilian, industrial, and military areas. A wireless sensor network (WSN) typically performs large-scale data collection with the help of low-power electronic devices. These devices, called sensors, are deployed with high density in a region of interest. The sensor nodes cooperatively form a multi-hop communication network over a wireless medium to accomplish common goals such as sensing, monitoring, and tracking. Sensors are empowered with sensing, processing, and communicating capabilities. In order to carry out these operations, a sensor node largely relies on the battery power embedded within its architectural framework.

In most WSN applications, the deployment region is inaccessible to humans. Hence, it becomes difficult to recharge the power source of sensor nodes. This requires sensor energy to be used in a highly efficient way so as to maximize the functional lifetime of the sensor network. A voluminous amount of research has been carried out on designing energy-efficient mechanisms so that the sensor activities can be optimized. Developers aim at building network protocols to conserve energy, allowing the network to operate without supervision for relatively longer periods of time. The unique sensor characteristics and network constraints presented in the WSN domain reveal new challenges for developing a wide range of technological innovations.

Our book is categorized into seven chapters to deal with graduate-level and research-oriented lecturing. Chapter 1 is drafted with the purpose of providing a complete systematic introduction to the fundamental concepts of sensor networks. It presents an overview of the low-power sensor, the network standards, design challenges, and sensor network simulation. Enhanced versions of energy models are also discussed and are enriched with energy harvesting.Chapter 2 reflects the wide range of sensor network applications, along with their integration into social networks and neural computing. Chapters 3 and 4 provide in-depth insights into coverage and connectivity issues with the help of genetic metaheuristics, evolutionary models, and various reinforcement methodologies designed for WSNs.Chapter 5 is focused on clustering and presents details about the various methods available for WSNs to tackle the energy hole problem and achieve load balancing and network lifetime enhancements. Chapter 6 illustrates the role of routing in sensor networks to achieve energy conservation. It discusses the classification, limitations, and advantages associated with routing methods. Chapter 7 presents a historical overview of performance evaluation methods along with the introduction of new performance metrics. After the chapters, we have two appendix sections to demonstrate network simulation and provide some additional materials for further studies.

Target audience. To summarize, the book addresses the prevailing methods, current trends, and future projections of research in sensor networks. Distinguished from other textbooks, this book presents in-depth analytical modelling that helps to foster an understanding of formidable network challenges in developing protocols for next-generation communication standards. Our textbook is designed with the aim of serving as a leading reference for those who seek broad knowledge with in-depth conceptualization. The target audience widely covers graduates, engineers, scientists, researchers, and industrial practitioners. Readers having initial knowledge of wireless networking as a prerequisite will easily grasp the contents.

Authors:
Dr. Adwitiya Sinha, Dr. Manju,
Dr. Samayveer Singh

MATLAB® is a registered trademark of The MathWorks, Inc. For product information, please contact:

The MathWorks, Inc.
3 Apple Hill Drive
MNatick, MA 01760-2098 USA
Tel: 508 647 7000
Fax: 508-647-7001
E-mail: info@mathworks.com
Web: www.mathworks.com

Author Biography

Dr. Adwitiya Sinha is working as Associate Professor in the Department of Natural and Applied Sciences, TERI School of Advanced Studies, New Delhi, India, since February 2024. Previously, she worked in academia and research for approximately 11 years in the Department of Computer Science and Engineering & Information Technology, Jaypee Institute of Information Technology, Noida-62, U.P., beginning in February 2013. Moreover, during her doctoral research in Jawaharlal Nehru University, New Delhi, India, she received the First Rank Certificate for excelling in the M.Tech. in year 2010. She was also awarded with a Senior Research Fellowship (SRF) by the Council of Scientific & Industrial Research (CSIR) in 2012. She has more than 90 publications in the form of SCI/Scopus journal articles, international conferences, books, and chapters. She has been in the Publication and Review Committee of the IEEE/ACM International Conference on Contemporary Computing since 2013. She has also delivered a lecture series on Networks & Graphs, Wireless Sensor Networks, and Performance Analysis of Computing Systems, organized by UGC, CEC, New Delhi in the form of EDUSAT live lectures. She has guided 21 post-graduate dissertation/industrial theses and more than 75 graduate minor/major projects in data science, social networking, large-scale graph algorithms, sensor networks, data mining, and machine learning. She is also supervisor to five Ph.D. scholars in the field of data science, social networking, and sensor networks, of which four have been awarded. For her contribution to research, she was promoted to IEEE Senior Member in 2019 by the Institute of Electrical & Electronics Engineers (IEEE), New York, USA. She is also an active professional member of the ACM research community.

Affiliation: Department of Natural and Applied Sciences, TERI School of Advanced Studies, Vasant Kunj, New Delhi, India

Dr. Manju received a Ph.D. in computer engineering from Delhi University, New Delhi, India, in 2018. She received a M.S. in computer and communication engineering from LNMIIT, Jaipur, India, in 2009 and a B.E. in computer engineering from Govt. Engineering College Kota, India, in 2006. She has published more than 30 times in reputed international journals and conferences. Presently, she is working as Assistant Professor in the Department of Computer Science & Engineering at Jaypee Institute of Information Technology (JIIT), Noida, India. Her research interests focus on algorithm designing, ad hoc networks, Internet of Things networks, and security in fog computing/edge.

Affiliation: Department of Computer Science and Engineering & Information Technology, Jaypee Institute of Information Technology, Noida, Uttar Pradesh, India

Dr. Samayveer Singh received his Ph.D. from the Department of Computer Engineering, Delhi University, New Delhi, India, in 2016. He obtained his M.Tech. in computer science and engineering from the National Institute of Technology, Jalandhar, Punjab, India, in 2010, and his bachelor's degree in information technology from Uttar Pradesh Technical University, Lucknow, India, in 2007. He has published more than 60 research papers in international journals and conferences. Currently, he is working as Assistant Professor in the Computer Science and Engineering Department, National Institute of Technology Jalandhar, Punjab, India. His research interests include wireless sensor networks, data hiding, and information security.

Affiliation: Department of Computer Science and Engineering, National Institute of Technology Jalandhar, Punjab, India

Chapter 1

SUMMARY

Wireless sensor networks (WSNs) have acquired significance owing to their wide range of applications in diverse environments. This chapter presents an in-depth exploration of WSNs, with a focus on low-power sensors, communication architectures, design challenges, energy management, and data-aggregation techniques. Low-power sensors play a crucial role in WSNs by enabling prolonged network operation and reducing the need for frequent battery replacement. The chapter discusses various aspects of low-power sensor design, including energy-efficient hardware, power management strategies, and sleep/wake scheduling mechanisms to maximize sensor lifespan while maintaining reliable operation. Moreover, the architecture of WSNs is examined, encompassing sensor nodes, base stations, and communication protocols. Different communication structures, such as point-to-point, multipoint-to-point, and multipoint-to-multipoint, are analyzed for their suitability in WSN deployments. Additionally, deployment strategies, including random, deterministic, and hybrid approaches, are discussed to optimize coverage and connectivity in WSNs. The chapter also focuses on wireless network standards relevant to WSNs, such as IEEE 802.15.4, Zigbee, and LoRa, highlighting their characteristics and applicability in different scenarios. Design challenges in WSNs are addressed, including energy conservation, self-organization, connectivity issues, energy holes, coverage problems, and routing challenges, offering insights into mitigation strategies and research directions. Besides this, energy management techniques are crucial for prolonging network lifetime and ensuring sustainable operation. The chapter discusses energy harvesting, adaptive transmission power control, and duty cycling methods to optimize energy consumption in WSNs.

Data aggregation is essential for reducing communication overhead and conserving energy in WSNs. Design issues, challenges, and the classification of data-aggregation techniques are also examined, thereby emphasizing their significance and scope in enhancing network efficiency and scalability. In conclusion, this chapter provides a detailed overview of advancements and challenges in WSNs, offering valuable insights for researchers, learners, and stakeholders interested in the design and development of wireless sensor networks for various applications.

DOI: 10.1201/9781003427780-1

1 Sensor Networks Overview

1.1 LOW-POWER SENSORS

Nowadays, a sensor network is generally considered a communication paradigm where a group of specified task-oriented devices known as sensors are used to monitor/record and respond to certain substances [1]. Depending on the requirements, these sensors can be helpful to monitor the temperature of given location, air pressure at a certain point, humidity of surroundings, wind direction and speed in a particular region, vibration intensity near an area, sound intensity inside a closed building, chemical concentrations inside a chemical factory, pollutant levels in a given atmosphere, and many more [2]. In order to do any of these tasks, a sensor node has to be configured with respect to the specific task. To perform this activity, one must know the basic architecture of the sensing device in detail so that the respective configuration can take place without any design fail. A network consists of tiny sensing devices that are densely deployed in a given proximity. There are six major components of a sensor device, namely, power source, sensing device, processor, transceiver, memory, and analog-to-digital convertor (ADC), as shown in Figure 1.1 [3]. It may consist of a few other application-specific components, such as a position finding system (e.g. GPS) and a mobilizer. In essence, the sensing component is used to sense the environment in terms of an analog signal and converts the received signal into digital form using an ADC unit. These signals are then passed to the processing unit so that the desired task can be performed. In order to store the data generated by the processor of a sensor, a memory unit is provided, which can store limited data for a certain amount of time. The most important and critical component of the sensing device is a power source, which is intended to be utilized in an energy-efficient manner, as it is non-rechargeable. Thus, this component has to be designed and placed in an efficient way to maximize the network's functional duration. Many applications of sensor networks need location information to perform certain tasks; therefore, a node may have a Global Positioning System (GPS) for location-related services. In order to support mobility for some application requirements, a mobilizer may sometimes be needed.

The major task of these sensing devices is to collect data from the deployed area. Further, the task is to transfer the collected data to the central unit, known as the base station (BS), where decisions are to be made. There are many general-purpose tasks where we need to use the functionality of various types of sensors as required by particular objectives [4, 5]. Based on the many sensor network applications, the

DOI: 10.1201/9781003427780-2

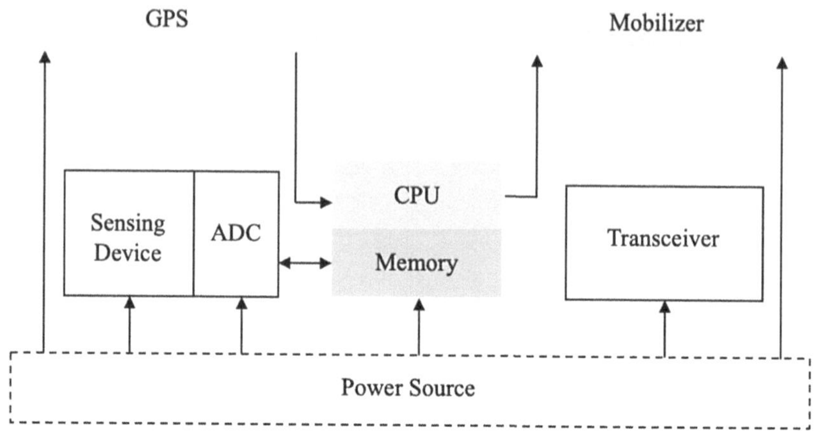

FIGURE 1.1   Sensor node architecture

following sections describe the types of sensor nodes that are mainly used by any general-purpose application based on sensors.

1.1.1 TEMPERATURE SENSOR

This is an analog type of sensor which measures the surrounding temperature. The output value is quite small, in the range of micro or mini volts. Due to this small-value measurement, an ADC unit is required in most applications. These sensors generally sense the change in the temperature in nearby surroundings.

1.1.2 PROXIMITY SENSOR

This type of sensor is generally a non-contact type of device which is mainly used for object detection in the near proximity. The sensor does not have to be in physical contact with the object to do this. Instead, objects' and sensors' Euclidean distances are calculated, and if they intersect, the proximity sensor can calculate this.

1.1.3 INFRARED SENSOR

The infrared (IR) sensor is a light-based sensor which is generally used by long-distance-based applications where objects are detected. These types of sensing devices are mainly used in mobile phones.

1.1.4 ULTRASONIC SENSOR

An ultrasonic sensor is a non-contact type of sensor which essentially measures the distance as well as the velocity of any object. These types of sensors are based on the sound waves' properties where working frequency is considerably higher than the human audible range. Thus, with ultrasonic sensors, one can easily measure the distance of various distant objects without requiring significant hardware.

1.1.5 Smoke and Gas Sensors

These sensors are very important where safety is the main concern of the given application. Most commercial buildings and industries have these devices installed in order to detect smoke and to further initiate an alarm device.

1.1.6 Alcohol Sensor

These sensors are used to detect alcohol levels. Generally, alcohol sensors are used in breathalyzer devices. As we know, these devices are used to measure a person's level of intoxication.

1.1.7 Humidity Sensor

Humidity sensors are used by weather monitoring systems to measure soil or atmosphere humidity levels. As we know, there are many farming requirements where measuring soil humidity along with temperature is important.

The sensor nodes described in the sections above are used in many important services offered by various organizations to the public. Next, we are going to discuss sensor network architecture in detail to understand its functionality.

1.2 WIRELESS SENSOR NETWORKS

The networks used for communication have changed and evolved tremendously over the years, and now they are entirely wireless rather than the earlier wired networks. Mobile phones, laptops, and cellular telephones are good examples of wireless communications. Wireless networks are categorized as ad hoc networks, mesh networks, opportunistic networks, cellular networks, peer-to-peer networks, wireless local area networks, and sensor networks [6, 7]. Based on functionality and infrastructure, wireless networks can be categorized as infrastructure-based networks and infrastructure-less networks [8, 9]. There is a predefined deployed communication network in infrastructure-based networks as there is in cellular networks. This type of network consists of networked devices, and to operate these devices, there are wireless access points. The access points are used to connect each device with other devices on the same network so that they can communicate with each other. The infrastructure-less network have no specific structure for devices such as ad hoc networks, wireless sensor networks (WSNs) with mobility, and static networks.

1.2.1 Sensor Network Architecture

In order to monitor a given area for specific tasks, a sensor network is deployed. The overall functionality of the sensor network is shown in Figure1.2. It is clearly depicted in the figure that sensing devices are closely deployed in random order in specific regions of interest (RoI) where they can gather data/information based on the type of node deployed for a particular application. The gathered information/data over the period is then seamlessly transferred to the central unit, or BS, either directly or via the neighbouring nodes for further processing, which includes removal of redundant data or irrelevant data. If the network is dense, the information is directly forwarded to the BS, but in the case of a sparse network where nodes are scattered far from each other, direct connection to the BS is almost impossible. Further, through satellite communication, the processed data is provided to end users for their personal use.

1.2.2 COMMUNICATION STRUCTURE IN WSNs

The wireless sensor network follows a protocol stack as shown in Figure 1.3, which is mainly used by the BS node and the remaining participating sensing devices deployed in the network for any type of communication. In order to compete this task, there are five major layers: the application layer, physical layer, network layer, data link layer, and transport layer [10]. These collectively are actors for the complete functionality of the underlying network paradigm. Each layer has a specific list of tasks which are performed from time to time at certain stages of the communication process while transferring information from one node to another and to the BS. In the following sections, we explain the core functionality of these layers.

1.2.2.1 Application Layer

The application layer mainly contains numerous application layer protocols that help in various sensor network applications. The main tasks of this layer are node localization, time synchronization, network security, and query dissemination. There are

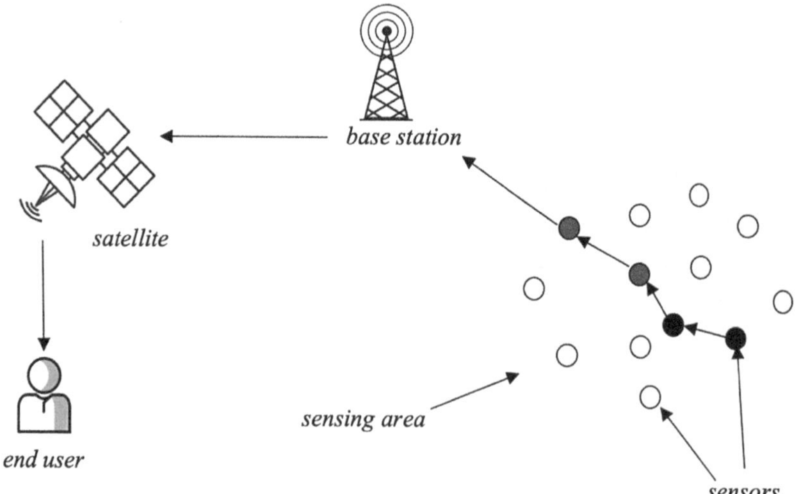

FIGURE 1.2 Wireless sensor network architecture

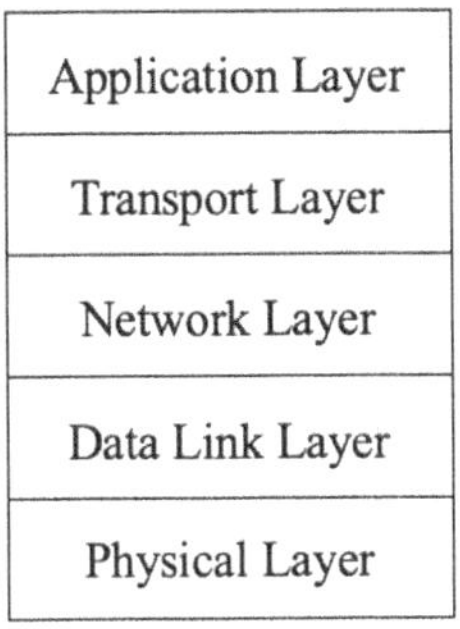

Application Layer
Transport Layer
Network Layer
Data Link Layer
Physical Layer

FIGURE 1.3 Communication protocol stack of sensor network

predominantly two protocols in this layer: the sensor management protocol (SMP), and the sensor query and tasking language (SQTL) [11]. The SMP protocol performs a variety of tasks such as node localization, querying a sensor node for status information, and scheduling sensor nodes.

1.2.2.2 Transport Layer

As we discussed earlier, there is continuous data transfer between nodes. The role of this type of transfer is managed by the transport layer, which ensures reliable data delivery. Because there are many critical constraints in WSN nodes, mainly including energy, storage, and computation, not all of the traditional transport protocols can be applied directly to WSNs. For example, the regular transport control protocol (TCP) cannot be used directly. In fact, the data delivery that takes place in wireless sensor networks primarily occurs in upstream and downstream flows. When sending information packets to the BS, the sensor nodes transmit their sensed data in a different way. Whenever the BS sends information packets to the sensors based on a certain process, it uses downstream transmission, which in turn requires different reliability measures [12].

1.2.2.3 Network Layer

In deployed wireless networks, this layer is mainly in use to route the data (sensed over the period) to the BS. To do this, these sensing devices send the data, including by single-hop or multi-hop transfer, to the BS. If the network is dense and nodes are quite close to the station, they can directly send it; this is known as single-hop transfer. However, in a sparse network, nodes are scattered over a huge area, and they need a few other nodes that act as bridging nodes to send the data to BS; this is known as multi-hop transfer.

1.2.2.4 Data Link Layer

The data link layer is responsible for data stream multiplexing, data frame creation and detection, medium access, and error control for reliable transmission. When performing these tasks, the medium access control (MAC) protocol is used. This protocol is more modified than the conventional network, as the wireless sensor network has some other constraints when it comes to infrastructure. Therefore, the modified MAC protocol is sufficiently capable of providing efficient shared communication resources in the given networks to improve network efficacy by reducing energy consumption and delivery latency.

1.2.2.5 Physical Layer

The main task performed by this layer is to convert bit streams that come directly from the layer situated above it into signals that are more suitable for transmission over the channel. Apart from signal transformation, this layer is also involved in dealing with the underlying hardware designs.

1.2.2.6 Management Planes

As well as the five layers of communication, the protocol stack also has a group of management planes spread over the layers. There are three management planes: power, connection, and task [13, 14]. The main role of a power management control is to manage the energy utilization of the sensing devices that are participating in certain task. Sensor nodes utilize power while sensing the environment, processing the data locally, transmitting this to the BS, and receiving the data back from the BS. For all these power-critical tasks, the role of the power management plane is to implement an energy-efficient mechanism so that power can be conserved and utilized in an efficient way. To do this, the MAC layer protocol can switch off nodes when it is idle in the given field. Additionally, at the network layer, when transmitting data to the BS, the sensor can choose a neighbour sensor that has the highest remaining energy for multi-hop communication. Essentially, the connection management plane plays an important role in providing configuration. It also plays an important role in maintaining connectivity. This task becomes crucial when there is a node failure due to certain conditions, and sometimes node movement or node displacement etc. is required.

1.2.3 DEPLOYMENT STRATEGIES IN WIRELESS SENSOR NETWORKS

The deployment strategies in sensor networks include deterministic deployment and random deployment, respectively [15]. In deterministic deployment, sensors are placed on pre-fixed locations. This kind of scheme is useful for those applications where human access is feasible in the network field, such as in-house activity monitoring, soil monitoring, and city sensors for urban monitoring. For such applications, sensor locations are first decided, then nodes are deployed in these locations only. In the case of random deployment, sensor nodes are usually spread over the region in random locations. This kind of deployment is preferable where the network area is humanly inaccessible, such as in fire detection, bird observation on Great Duck Island, or other wildlife monitoring. Figure 1.4 shows deployment methods where the network area is fixed, and sensor nodes are deployed under deterministic and non-deterministic scenarios.

There is no particular rule to decide on the deployment schemes. Depending on the type of application, the respective deployment paradigm is chosen so that node

FIGURE 1.4 Deployment strategies: (a) deterministic and (b) non-deterministic

deployment tasks can be smoothly implemented. Nowadays, most of the applications of WSNs are in terrains where human access is either limited or impossible. Thus, random deployment is most frequently used.

1.3 WIRELESS NETWORK STANDARDS

Once the deployed network ensures that all the sensors are connected to the relevant neighbouring nodes under a certain topology, communication between these nodes becomes the next challenge. To communicate, these connected sensors follow a particular communication paradigm based on the requirement fixed by the underlying network architecture and the application. In the literature, there are many communication protocols that can be adapted by the deployed network to enable communication between the participating nodes [16].

These communication technologies are mainly classified as short-distance wireless communication technology, medium-distance wireless communication technology, and long-distance communication technology [17]. Radio frequency identification (RFID) and Bluetooth are short-distance protocols; Wi-Fi [18] and Zigbee [19] are medium-distance protocols; and cellular networks (2G/3G/4G) and low-power, wide-area (LPWA) are mainly used for long-distance communication protocols. In Chapter 4, we will discuss these communication protocols in detail for better understanding, where the compete functionality is well described.

1.4 DESIGN CHALLENGES

The wireless sensor and control methodologies are primarily utilized to narrow the gap between the physical world designed for humans and the virtual world designed by machines. Wireless sensor networks hold the potential to provide a low-cost solution for daily problems, including home automation and building surveillance. However, owing to limited storage and the energy of sensor nodes, many other issues and challenges are faced when setting up a functional wireless network. Some of the prominent design issues are limited energy, self-localization of sensing nodes, connectivity issues between the nodes, energy holes formed over a period, coverage issues, and routing challenges when transferring information from one node to another in the network or the BS. Next, we will discuss these issues one by one, in detail, to understand the major obstacles.

1.4.1 Energy Conservation

We have already described that sensor nodes are tiny devices and that they are equipped with limited power sources because the size of the nodes limits the size of the battery. There are many wireless sensor network applications used in wildlife monitoring, the chemical industry, and military areas, where network terrain is not frequently accessible due to various security concerns. In such scenarios, sensor batteries cannot be replaced or recharged once they have become exhausted. Due to this, one has to utilize the sensor batteries in efficient and optimal ways so that network lifetime can be extended for a longer period. In the literature, researchers have proposed many energy-efficient, energy-conserving protocols to maximize network lifetime in various applications of sensor networks, including area coverage [20], target coverage [21], connectivity in network [22], and clustering methods [23]. All the methodologies addressed are mainly focused on conserving node energy for longer periods so that the network can be functional for extended durations.

1.4.2 Self-Organization and Clustering

Because network nodes are deployed in a random fashion most of the time, every node should be capable of localizing itself without any human intervention. Each deployed node should be able to manage network configuration, adaptation, maintenance, and repair by itself [24]. This is challenging to achieve because there are situations where nodes are not able to manage themselves. There are many applications where the central node (i.e. BS) has to collect data from the deployed network nodes at the time of making any further decisions. To send the data collectively to the BS, there are two possibilities for sensors. Sensors can either send the data individually or in a group of neighbouring nodes. For example, when one has to measure the humidity of a given area, then we know that nearby sensors have almost the same humidity level measurements. Therefore, in such situations, instead of sending individual pieces of data to the BS, clusters are formed to send collective readings. Finding clusters in a such a way that there is no redundancy of data at the BS is another challenge. Many researchers have addressed this issue and proposed various energy-efficient protocols to form clusters to minimize battery usage.

1.4.3 Wireless Domain and Connectivity Issues

Sensor nodes are scattered in the deployed wireless sensor networks where there is no physical or wired connection between the sensing devices. We also know that sensors are deployed to sense the surroundings based on the type of application. So, once data is ready at any sensing device, the collected data has to be sent to the BS where further decisions can be made. In this regard, the sensing devices need to be connected to the BS using either single-hop if they are adjacent to the BS, or using multi-hop if they are further from the BS. Achieving connectivity is the biggest challenge for any type of connection, as these devices are wirelessly deployed [25]. Merely providing connectivity is not the prime objective; instead, energy-efficient connectivity is the main concern in a power-scarce network. Researchers have, therefore, thoroughly addressed this issue and have come up with energy-efficient protocols for various types of application where continuous connectivity is required throughout the network's functional duration [26].

1.4.4 ENERGY HOLES AND COVERAGE ISSUES

In a sensor network, energy holes refer to a wireless network scenario where certain sensor nodes exhaust their energy resources much rapidly than other nodes. This non-uniform depletion of energy creates disconnected areas in the network, also known as energy holes, where sensors are unable to communicate and function effectively due to energy exhaustion [27]. There exist several techniques to alleviate the energy hole problem in sensor networks. These include optimal deployment, selection of routing protocol, data aggregation, energy harvesting, and duty scheduling with load balancing [28].

The optimal deployment strategies of sensor nodes consider network coverage, connectivity, rate of energy consumption, and residual energy of the network [29]. An optimal method of sensor deployment reduces the chances of energy holes and maximizes the network performance. There are different deployment strategies, such as random, deterministic, uniform, and hierarchical. Sensor nodes can be randomly dispersed across the deployment area, or deterministically deployed with careful planning based on the characteristics of the deployment area. In addition, the distribution can be uniform, or may be deployed into multiple layers to reduce energy consumption and improve scalability. Moreover, the suitable choice of low-energy routing protocols can help to balance usage and network coverage [30]. This involves dynamically selecting paths, which eventually minimizes energy expenditure while relaying the data packets. Another way of reducing energy holes in the network includes data-aggregating methods. Aggregation involves the processing of data at intermediate nodes prior to transmission. This effectively reduces the number of data transmissions and retransmissions, thereby saving energy. Section 1.6 provides more details on aggregation.

The WSN is typically comprised of battery-based sensor electronics that are distributed in a wide geographical region to collect and transmit data. Hence, for managing the energy of an individual node, sleep and wake cycles can be scheduled. This allows dynamic load balancing, which leads to conservation of energy during periods of inactivity, thus preventing premature energy depletion and ensuring continuous operation.

1.4.5 ROUTING CHALLENGES IN SENSOR NETWORKS

Sensor networks can face numerous difficulties due to resource limitations inherent in the systems. Sensor nodes typically operate with constrained resources, including limited energy, processing capabilities, and communication range. This requires the routing algorithms to give precedence to pathways that reduce energy consumption, implement duty scheduling, and integrate strategies that are attuned to energy considerations [31]. Hence, routing mechanisms become imperative to guarantee the sustained functionality of the network. Sensor networks are often deployed in dynamic environments, where the network topology undergoes frequent changes. The dynamic nature of the topology is influenced by factors such as node movements, fluctuations in environmental conditions, and alterations in communication patterns. This requires the routing protocols to exhibit adaptability for ensuring continual network connectivity and operation [32, 33]. This adaptability is crucial for maintaining seamless functionality despite the inherent fluctuations in node positions and connectivity within the dynamic environment. Moreover, the sensor nodes are engaged in the continuous generation and transmission of data according to how critical the application is. This requires efficient routing strategies to incorporate mechanisms for aggregating and processing data at intermediate nodes before forwarding it to the BS [34]. This process of data aggregation serves to diminish the overall volume of transmitted data, concurrently minimizing energy consumption and optimizing bandwidth utilization.

Several researchers are actively exploring inventive algorithms and protocols, seeking a harmonious equilibrium between energy efficiency, scalability, fault tolerance, data aggregation, and adaptability, and dynamic network conditions. This helps in elevating the overall performance and reliability of sensor networks across diverse applications, encompassing areas such as environmental monitoring, smart cities, smart healthcare, industrial automation, social sensing, and intelligent farming.

1.5 ENERGY MANAGEMENT

Energy is a critical aspect in sensor networks, and hence it should be managed carefully to ensure a long functional lifetime [35]. This is because most sensor-based applications require sensor deployment in remote or inaccessible locations where battery replacement or recharging is expensive or difficult. The energy consumption of the sensors in resource-constrained environments can be managed in several ways, which primarily include optimizing duty–sleep cycles and utilizing low-power hardware electronics and software protocols.

The first order radio model is the benchmark technique to compute energy dissipation during transmission and reception activities. Each sensor node has two modules for communication [36]. The first module is the transmitter, which dissipates energy to run the radio electronics and the power amplifier unit. The second module is the receiver, which only expends sensor energy to run the radio electronics unit. If d_o denotes the threshold distance, then two types of power control level can be used by appropriately setting the power amplifier, which includes the following modes:

If $d < d_o$, then the free space model is used.

If $d > d_o$, then the multipath model is preferred.

To transmit an l-bit message, at distance d, the energy consumption in the transmission and reception modules can be expressed as in Equations (1–3).

$$E_{Tx}\left(l,d\right)=E_{Tx-elec}\left(l\right)+E_{Tx-amp}\left(l,d\right) \tag{1.1}$$

$$E_{Tx}\left(l,d\right)=\begin{cases} lE_{elec}+l\epsilon_{fs}d^2 & d<d_o \\ lE_{elec}+l\epsilon_{mp}d^4 & d\geq d_o \end{cases} \tag{1.2}$$

$$E_{Rx}\left(l\right)=E_{Rx-elec}\left(l\right)=lE_{elec} \tag{1.3}$$

In the above equations, E_{elec} denotes the radio electronics energy constant that amounts to $50\ nJ\,/\,bit$. Moreover, there are amplifier energy constants, namely $\epsilon_{fs}=10\ pJ\,/\,bit\,/\,m^2$ and $\epsilon_{mp}=0.0013\ pJ\,/\,bit\,/\,m^4$, respectively.

1.6 DATA AGGREGATION

In a wireless sensor network, minimizing energy consumption is the primary challenge for increasing the network lifetime. The sensors are distributed in a huge quantity for continuous extraction of data from the network. These sensors communicate over wireless channels and cooperate with each other to accomplish a fusion of sensory data [37]. A huge amount of sensed data, produced over a span of time, is highly correlated and redundant. The transmission of such a massive amount of data would deplete sensor energy at a very rapid rate. Colossal amounts of data accumulated at the sink require computational overhead while processing such data. Therefore, efficient aggregation techniques are important strategies to conserve energy in power-constrained sensor networks [38].

1.6.1 DESIGN ISSUES AND CHALLENGES

Developing efficient aggregation schemes for managing sensor data is inherently a challenging task [39]. Some of the major design goals are as follows:

a. Increasing energy efficiency: Energy is considered a prime resource for battery-powered sensor nodes. Hence, an effective data-aggregation scheme should aim to minimize the energy consumption of all the sensors in the network with the intention that each sensor should consume approximately the same amount of energy in every data-aggregation cycle.

b. Enhancing network lifetime: The timespan during which the sensor network remains operational is another performance evaluation criterion that helps in measuring the efficiency of a sensor network in terms of energy consumption. Network lifetime can be elaborated as total data-aggregation rounds successfully completed with at least x fraction of sensors remaining active in the network. Here, x refers to the fraction of nodes in the absence of which the network stops functioning. Data-aggregation techniques should be devised in such a manner that energy drainage remains uniform throughout the network.

c. Maintaining data accuracy: While performing aggregation, a certain amount of data is filtered out due to redundancy, errors, or insufficiency of storage, which leads to data unreliability. Thus, the proposed aggregation scheme should consider such aspects so that the aggregated information is accurate enough to convey correct information to the sink.

d. Minimizing latency: The total amount of delay incurred in processing, aggregating, and transmitting sensory data to its intended destination is known as latency. Latency can be further described as the time measured between data recorded by an individual source node and data reception at the sink. An efficient aggregation scheme should attempt to minimize computational and communicational latency.

Therefore, the design of data-aggregation mechanisms should mainly focus on improving energy efficiency, network lifetime, and data accuracy with minimal communication and computational latency.

1.6.2 SIGNIFICANCE AND SCOPE

The amount of data generated by a WSN with a large number of sensors is typically too large to be managed by the BS. Because sensors are power-constrained devices, it is considered wasteful for each sensor node to communicate its entire dataset directly to the BS [40]. Moreover, the tendency for datasets to be gathered by neighbouring sensors is usually correlated in space and time, i.e. spatio-temporal redundancy. Therefore, it is essential to develop aggregation methods to combine large sets of data into concise pieces of information at intermediate sensor nodes. This minimizes data flows sent to the BS, leading to productive conservation of energy and bandwidth.

Data aggregation refers to the process of summarizing samples gathered from a collection of neighbouring sensors before reporting to the BS for further processing [41]. Aggregation allows the extraction of raw data from sensor nodes and makes the processed version available at the BS with tolerable latency. Data aggregation is considered one of the most important factors that determines the degree to which efficiency of a sensor network can be exploited with minimum energy consumption. Data-aggregation methods greatly help sensor networks to function by:

- Collaborative mining of received information.
- Sufficiently compressing sensor traffic prior to transmission.
- Enhancing quality of fused data (i.e. drawing a collective conclusion).
- Minimizing latency involved in sensor communication.
- Avoiding network congestion due to redundant transmission.
- Utilizing network bandwidth optimally.
- Conserving sensor energy to enhance operational lifetime.
- Improving reliability and network throughput to a large extent.

Data-aggregation techniques are highly application-specific and infrastructure-dependent. Previous studies have shown several promising and relevant contributions in related areas with significant scope for future research.

1.6.3 DATA-AGGREGATION CLASSIFICATION

The process of data aggregation is crucial to sensor networks and results in reduced data traffic that ultimately minimizes energy consumption in the network. The aggregation methods in WSNs can be further classified based on their functionalities and approaches. These include categorizations based on the network architecture, data flows in the network, and the quality of service requirements. The following section highlights each of these classifications and its significance to specific sensor network scenarios.

1.6.3.1 Network Architecture-Based Classification

In general, two types of communication networks can be formed: flat networks and hierarchical networks [42]. In flat networks, all the sensors are considered homogeneous in terms of functionality, i.e. each node is equipped with the same battery power to sense, aggregate, and transmit data to the sink. In such types of network, aggregation is performed with the help of data-centric routing, which involves flooding of a query by the BS (or sink) to all sensor nodes, and those possessing the required data respond back within a predefined time frame. The selection of communication protocol and routing strategy depends on the specific requirements of the end-user application. Direct diffusion (push/pull diffusion) is one of the most commonly used protocols in flat sensor network platforms. The burden of communication and computation tends to rapidly increase at the sink with an increasing number of sensors in the WSN and results in a rapid depletion of sensor power. Hence, to avoid breakdown in the functionality of the entire network, a hierarchical data-aggregation approach is used. This hierarchical aggregation scheme allows data fusion to take place at specific nodes in order to reduce bulk transmissions to the BS. This eventually improves the energy utilization and lifetime efficiency of the wireless sensor network. Some significant hierarchical data-aggregation strategies are illustrated as follows:

- Cluster-based network: In a power-constrained network with a large number of sensors, it is highly inefficient to transmit complete sets of data to the BS. In a clustered scenario, each node belongs to a specific cluster. Every cluster possesses a cluster head that acts as a local aggregator. It processes the data gathered from each sensor node within its cluster boundary and transmits the fused information to the sink, which results in significant conservation of sensor energy. Some well-known protocols belonging to this category are the low-energy adaptive clustering hierarchy (LEACH) [43] and the hybrid energy-efficient distributed clustering approach (HEED) [44].
- Chain-based network: In a cluster-based wireless scenario, the cluster heads are assigned the task of data aggregation. However, if a head node

is situated at the cluster border, most of the packets from the sensor nodes have to traverse long distances to be delivered to the sink, which might excessively deplete sensor energy. In such cases, energy efficiency can be obtained by transmitting data to close neighbours. This is achieved by chain-based data aggregation, where the main idea is to allow sensors to transmit only to their closest neighbouring node. The efficiency of the chain-based data-aggregation scheme is largely governed by the creation of efficient data chains, which ultimately minimize energy consumption. Certain factors such as network density and sink location impact the choice of an appropriate chain construction algorithm. The power-efficient data-gathering protocol for sensor information systems (PEGASIS) is based on the principles of a chain-based aggregation scheme [45].

- Tree-based network: A tree-based sensor network organizes the sensors into a spanning tree in which aggregation is performed on the data generated by intermediate nodes existing towards the sink situated at the root. This method is widely known as the in-network aggregation of sensory data. Applications such as monitoring radiation levels in nuclear plants, in which data is provided by each node, is considered very useful for ensuring safety. Constructing an energy-efficient data-aggregation tree is considered an important aspect for improving the performance of tree-based networks. In-network sensor data aggregation is suitable for environments where the events are greatly correlated in space and time.

- Grid-based network: In grid architectures, a sensor network is sub-divided into fixed regions (or grids). A grid aggregator node is set in each grid with the task of processing and aggregating data collected from nodes within the grid. Unlike cluster-based scenarios, a sensor within a grid prefers to directly communicate with the grid aggregator rather than interacting with the neighbouring sensor nodes. Grid architecture-based data aggregation is preferable for applications with random mobility such as military surveillance, which requires adaptability to dynamic network changes and object mobility.

1.6.3.2   Network Flow-Based Classification

Data-aggregation processes that can be modelled as network flow problems are classified under network flow-based schemes [46]. Flow-based protocols aim to optimize network lifetime under information flow constraints based on network and sensor energy constraints. This section illustrates two major flow-based approaches: maximum lifetime data aggregation and energy-constrained network-flow optimization.

- Maximum lifetime data aggregation: The maximization of network functionality over time is achieved using data aggregation that allows sensor nodes to aggregate multiple incoming data streams into a single high-information-bearing stream. The WSN is modelled in the form of a directed graph $G_1 = (V, E)$. Graph edges are associated with a function $f(u, v)$ that denotes the number of packets being sent from sensor u to v, respectively. An optimal solution signifying admissible network flow is obtained using linear integer programming. This maximizes the lifetime of network T_{G_1} under power and capacity constraints.

- Energy constrained network flow optimization: Data gathering can be modelled as a network flow optimization problem. The maximal data gathering approach operates with the aim of increasing the effective number of data collection cycles because of restricted power and other sensor resources. These constraints are further transformed into capacities associated with graph edges. This allows nodes to produce a fixed amount of data packets in every data-aggregation round. Hence, the problem of maximal data collection ultimately becomes reduced to addressing the problem of restricted flow problem with different edge capacities. This redefines the original problem as determining the existence of any data flow that complies with the data flow and link capacity constraints. This allows the network to be modelled as graph $G_2 = (V, E)$, with edge capacity denoted as $h(u, v) = \eta A_N$, where η is the total amount of packets generated in each aggregation round and A_N represents the overall number of such rounds.

1.6.3.3 Quality of Service-Based Classification

Several applications require aggregation mechanisms to meet a predefined level of quality of service in terms of utilization of network bandwidth, delay, and throughput [47]. This leads to a QoS-based classification that lays its main focus on ensuring QoS metrics with an anticipated performance measurement. Quality of service-aware data-aggregation strategies can be further divided into the following two categories:

- Optimal information extraction: This includes aggregation methods that focus on the amount of information reported to the BS under energy, delay, and flow constraints.
- End-to-end reliability and congestion control: This involves those aggregation methods that essentially emphasize the control of network congestion and improve end-to-end data reliability [48]. The main aim is to achieve maximum communication channel utilization.

Overall, data aggregation in wireless sensor communication reduces the total amount of outgoing traffic, which ultimately helps in reducing packet losses because of network collisions and route congestions [49]. Apart from this, aggregation techniques also improve the authenticity of aggregated information by reducing redundancy and other ambiguities in data.

REFERENCES

1. Akyildiz F., W. Su, Y. Sankarasubramaniam, and E. Cayirci. "A Survey on Sensor Networks." *IEEE Communications Magazine* 40, no. 8 (2002): 102–14.

2. Majid, Mamoona, et al. "Applications of Wireless Sensor Networks and Internet of Things Frameworks in the Industry Revolution 4.0: A Systematic Literature Review." *Sensors* 22, no. 6 (2022): 2087.

3. Buurman, Ben, et al. "Low-power Wide-area Networks: Design Goals, Architecture, Suitability to Use Cases and Research Challenges." *IEEE Access* 8 (2020): 17179–220.

4. Kandris, Dionisis, et al. "Applications of Wireless Sensor Networks: An Up-to-Date Survey." *Applied System Innovation* 3 no. 1 (2020): 14.

5. Sharma, Ravi, Shiva Prakash, and Pankaj Roy. "Methodology, Applications, and Challenges of WSN-IoT." 2020 International Conference on Electrical and Electronics Engineering (ICE3), IEEE, 2020.

6. Mini S. Udgata S.K., and Sabat S.L. "Sensor Deployment and Scheduling for Target Coverage Problem in Wireless Sensor Networks." *IEEE Sensors Journal* 14 no. 3, (2014): 636–44.

7. Pujari, A. K., S. Mini, and T. Padhi. "Polyhedral Approach for Lifetime Maximization of Target Coverage Problem." *ICDCN* (2015): 14:1–14:8. https://doi.org/10.1145/2684464.2684495.

8. M. Chaudhary and A. K. Pujari. "Q-Coverage Problem in Wireless Sensor Networks." In *Proc. Int. Conf. Distrib. Comput. Netw. ICDCNin (Lecture Notes in Computer Science)*, vol. 5408. Berlin, Germany: Springer-Verlag, 2009, pp. 325–30.

9. Manju, Pawan Bhambu, and Sandeep Kumar. "Target K-Coverage Problem in Wireless Sensor Networks." *Journal of Discrete Mathematical Sciences and Cryptography* 23 no. 2, (2020): 651–659.

10. Goyal, Nitin, Mayank Dave, and Anil Kumar Verma. "Protocol Stack of Underwater Wireless Sensor Network: Classical Approaches and New Trends." *Wireless Personal Communications* 104 (2019): 995–1022.

11. Manju, Chand S., and B. Kumar. "Selective α-Coverage Based Heuristic in Wireless Sensor Networks." *Wireless Pers Commun* 97 (2017): 1623–36.

12. Manju, Anuradha. "A Novel Energy-Efficient Heuristic for Target Coverage to Maximize Sensor Network Lifetime." *International Journal of Computer Applications (IJCA)* 86 no. 7, (January 2014): 31–5.

13. Hawbani, Ammar, et al. "Novel Architecture and Heuristic Algorithms for Software-Defined Wireless Sensor Networks." *IEEE/ACM Transactions on Networking* 28, no. 6 (2020): 2809–22.

14. Tan, Xiaobo, et al. "QSDN-WISE: A New QoS-based Routing Protocol for Software-Defined Wireless Sensor Networks." *IEEE Access* 7 (2019): 61070–82.

15. Manju, Samayveer Singh, Anshu Kumar Dwivedi, AK Sharma, Pawan Singh Mehra. "Learning Automata Based Heuristics for Target Q-Coverage." 2020 8th International Conference on Reliability, Infocom Technologies and Optimization (Trends and Future Directions) (ICRITO), pp. 170–3, 2020.

16. Mohamadi, H., A. S. Ismail, and S. Salleh. "Solving Target Coverage Problem Using Cover Sets in Wireless Sensor Networks Based on Learning Automata." *Wireless Personal Communications* 75 (2014): 447–63.

17. Beom-Su Kim, Ho Sung Park, Kyong Hoon Kim, Daniel Godfrey, and Ki-Il Kim. "A Survey on Real-Time Communications in Wireless Sensor Networks." (2017). https://doi.org/10.1109/icoin.2017.7899476.

18. Landaluce, H., L. Arjona, A. Perallos, F. Falcone, I. Angulo, and F. Muralter. "A Review of IoT Sensing Applications and Challenges Using RFID and Wireless Sensor Networks." *Sensor* 20, no. 9 (2020): 2495.

19. Meena Ahlawat, and Ankita Mittal. "Different Communication Protocols for Wireless Sensor Networks: A Review." *International Journal of Advanced Research in Computer and Communication Engineering* 4, no. 3 (March 2015). https://doi.org/10.17148/ijarcce.2015.4351.
20. Hanh, Nguyen Thi, et al. "An Efficient Genetic Algorithm for Maximizing Area Coverage in Wireless Sensor Networks." *Information Sciences* 488 (2019): 58–75.
21. Saadi, Nora, et al. "Maximum Lifetime Target Coverage in Wireless Sensor Networks." *Wireless Personal Communications* 111 (2020): 1525–43.
22. Ratha, Pradyumna Kumar, Siba K. Udgata, and Nihar Ranjan Satapathy. "A Sensor Deployment Scheme for Fault-Tolerant Connected Probabilistic Target Coverage: A Trade-Off Among Coverage, Connectivity, and Fault Tolerance." Intelligent Systems: Proceedings of ICMIB 2020, Springer Singapore, 2021.
23. Singh, Samayveer. "An Energy Aware Clustering and Data Gathering Technique Based on Nature Inspired Optimization in WSNs." *Peer-to-Peer Networking and Applications* 13 (2020): 1357–74.
24. Orfanus, Dalimir, Tales Heimfarth, and Peter Janacik. "An Approach for Systematic Design of Emergent Self-Organization in Wireless Sensor Networks." 2009 Computation World: Future Computing, Service Computation, Cognitive, Adaptive, Content, Patterns, IEEE, 2009.
25. Zhu, Chuan, Chunlin Zheng, Lei Shu, and Guangjie Han. "A Survey on Coverage and Connectivity Issues in Wireless Sensor Networks." *Journal of Network and Computer Applications* 35, no. 2 (2012): 619–32.
26. Ghosh, Amitabha, and Sajal K. Das. "Coverage and Connectivity Issues in Wireless Sensor Networks: A Survey." *Pervasive and Mobile Computing* 4, no. 3 (2008): 303–34.
27. Zhang, Yunzhou, Xiaohua Zhang, Wenyan Fu, Zeyu Wang, and Honglei Liu. "HDRE: Coverage Hole Detection with Residual Energy in Wireless Sensor Networks." Journal of Communications and Networks 16, no. 5 (2014): 493–501.
28. Mohemed, Reem E., Ahmed I. Saleh, Maher Abdelrazzak, and Ahmed S. Samra. "Energy-Efficient Routing Protocols for Solving Energy Hole Problem in Wireless Sensor Networks." *Computer Networks* 114 (2017): 51–66.
29. Ramos, Heitor S., Azzedine Boukerche, Alyson L. C. Oliveira, Alejandro C. Frery, Eduardo M. R. Oliveira, and Antonio A. F. Loureiro. "On the Deployment of Large-Scale Wireless Sensor Networks Considering the Energy Hole Problem." *Computer Networks* 110 (2016): 154–67.
30. Asharioun, Hadi, Hassan Asadollahi, Tat-Chee Wan, and Niyayesh Gharaei. "A Survey on Analytical Modeling and Mitigation Techniques for the Energy Hole Problem in Corona-Based Wireless Sensor Network." *Wireless Personal Communications* 81 (2015): 161–87.
31. Radi, Marjan, Behnam Dezfouli, Kamalrulnizam Abu Bakar, and Malrey Lee. "Multipath Routing in Wireless Sensor Networks: Survey and Research Challenges." *Sensors* 12, no. 1 (2012): 650–85.
32. Sharma, Ashutosh, Rajiv Kumar, and Pawandeep Kaur. "Study of Issues and Challenges of Different Routing Protocols in Wireless Sensor Network." In 2019 Fifth International Conference on Image Information Processing (ICIIP), pp. 585–90, IEEE, 2019.
33. Nayak, Padmalaya, G. K. Swetha, Surbhi Gupta, and K. Madhavi. "Routing in Wireless Sensor Networks Using Machine Learning Techniques: Challenges and Opportunities." *Measurement* 178 (2021): 108974.
34. Raja Basha, Adam. "A Review on Wireless Sensor Networks: Routing." *Wireless Personal Communications* 125, no. 1 (2022): 897–937.

35. Martinez-Sala, Alejandro, Jose-Maria Molina-Garcia-Pardo, Esteban Egea-Ldpez, Javier Vales-Alonso, Leandro Juan-Llacer, and Joan Garcia-Haro. "An Accurate Radio Channel Model for Wireless Sensor Networks Simulation." Journal of Communications and Networks 7, no. 4 (2005): 401–7.
36. Bello-Salau, H., A. F. Salami, F. Anwar, and Md Rafiqul Islam. "Analysis of Radio Model Performance for Clustering Sensor Networks." *Sensors & Transducers* 128, no. 5 (2011): 27.
37. Bagaa, Miloud, Yacine Challal, Adlen Ksentini, Abdelouahid Derhab, and Nadjib Badache. "Data Aggregation Scheduling Algorithms in Wireless Sensor Networks: Solutions and Challenges." *IEEE Communications Surveys & Tutorials* 16, no. 3 (2014): 1339–68.
38. Sinha, Adwitiya, Ratik Puri, Udit Balyan, Ritik Gupta, and Ayush Verma. "Sustainable Time Series Model for Vehicular Traffic Trends Prediction in Metropolitan Network." In 2020 6th International Conference on Signal Processing and Communication (ICSC), pp. 74–9, IEEE, 2020.
39. Cai, Simin, Barbara Gallina, Dag Nyström, and Cristina Seceleanu. "Data Aggregation Processes: A Survey, a Taxonomy, and Design Guidelines." *Computing* 101 (2019): 1397–429.
40. Sarangi, Kaustuv, and Indrajit Bhattacharya. "Utility of Data Aggregation Technique for Wireless Sensor Network: Detailed Survey Report." In Social Transformation–Digital Way: 52nd Annual Convention of the Computer Society of India, CSI 2017, Kolkata, India, January 19–21, 2018, Revised Selected Papers 52, pp. 68–82, Springer Singapore, 2018.
41. Thakur, Monica, Rajeev Bedi, and Dr Sunil Kumar Gupta. "A Review on Data Aggregation Techniques for Wireless Sensor Networks." In Proceedings of International Conference on Information Technology and Computer Science, July, pp. 11–2, 2015.
42. Randhawa, Sukhchandan, and Sushma Jain. "Data Aggregation in Wireless Sensor Networks: Previous Research, Current Status and Future Directions." *Wireless Personal Communications* 97 (2017): 3355–425.
43. Malik, Meena, Yudhvir Singh, and Anshu Arora. "Analysis of LEACH Protocol in Wireless Sensor Networks." *International Journal of Advanced Research in Computer Science and Software Engineering* 3, no. 2 (2013). https://doi.org/10.18535/ijecs/v5i11.08.
44. Younis, Ossama, and Sonia Fahmy. "HEED: A Hybrid, Energy-Efficient, Distributed Clustering Approach for Ad-hoc Sensor Networks." *IEEE Transactions on Mobile Computing* 3, no. 4 (2004): 366–79.
45. Lindsey, Stephanie, and Cauligi S. Raghavendra. "PEGASIS: Power-Efficient Gathering in Sensor Information Systems." In Proceedings, IEEE Aerospace Conference, vol. 3, p. 3, IEEE, 2002.
46. Song, Sui, Li Ling, and C. N. Manikopoulo. "Flow-Based Statistical Aggregation Schemes for Network Anomaly Detection." In 2006 IEEE International Conference on Networking, Sensing and Control, pp. 786–91, IEEE, 2006.
47. Jeong, Jongsoo, Jaeseok Kim, Woosuk Cha, Haeyong Kim, Sangcheol Kim, and Pyeongsoo Mah. "A QoS-Aware Data Aggregation in Wireless Sensor Networks." In 2010 The 12th International Conference on Advanced Communication Technology (ICACT), vol. 1, pp. 156–61, IEEE, 2010.
48. Dubey, Arpan Kumar, and Adwitiya Sinha. "Congestion Control for Self Similar Traffic in Wireless Sensor Network." In 2015 Eighth International Conference on Contemporary Computing (IC3), pp. 331–35, IEEE, 2015.
49. Zhu, Jin, Symeon Papavassiliou, and Jie Yang. "Adaptive Localized QoS-Constrained Data Aggregation and Processing in Distributed Sensor Networks." *IEEE Transactions on Parallel and Distributed Systems* 17, no. 9 (2006): 923–33.

Chapter 2

SUMMARY

Intelligent sensor networks have emerged as a pivotal technology revolutionizing various domains, from domestic settings to complex industrial operations. This chapter explores the multifaceted applications and the integration of intelligent sensor networks with social networks and neural computing paradigms. For smart homes, sensor networks enable automation and energy efficiency by monitoring environmental conditions and human activities. Similarly, in healthcare, these networks facilitate remote patient monitoring, early detection of health anomalies, and personalized care delivery. Also, agriculture benefits from intelligent farm sensing, optimizing resource utilization, crop monitoring, and enhancing yield. Terrestrial, underwater, and underground sensor networks enable environmental monitoring, disaster management, and resource exploration in challenging terrains. Intelligent vehicle sensors enhance road safety and traffic management through real-time data collection and analysis in intelligent transportation systems (ITS). Automation and control systems leverage sensor data for process optimization, predictive maintenance, and quality assurance. Moreover, military surveillance relies on sensor networks for situational awareness and threat detection. Furthermore, hyperspectral and satellite sensors offer advanced remote observation capabilities for environmental monitoring and space exploration. Also, integration with social networks introduces new dimensions to sensor networks, enabling collaborative sensing, data sharing, and crowd-sourced information gathering. Analysis of random graph formations, degree distributions, and node centrality enhances understanding and optimization of social-sensor networks. Social sensing applications harness the collective intelligence of networked individuals for various tasks, from disaster response to urban planning. Furthermore, integration with neural computing facilitates advanced data processing, pattern recognition, and decision-making in sensor networks. Neural networks enhance sensor data interpretation, anomaly detection, and adaptive control, leading to more intelligent and autonomous systems. This chapter provides a comprehensive overview of the state-of-the-art in intelligent sensor networks and explores their integration with social networks and neural computing, highlighting their transformative potential across diverse domains.

DOI: 10.1201/9781003427780-3

2 Sensor Network Applications

2.1 INTELLIGENT SENSOR NETWORKS

The advent of sensor-based technology in everyday applications has greatly improved efficiency and has brought advantages to data-driven decision-making practices [1]. Environmental monitoring is one of the benchmark applications of sensor networks, where the sensing devices are used for evaluating air quality, water purity, monitoring particulate matter, etc. [2, 3]. This sensory data can be further utilized for providing weather predictions, climate change insights, and also help in building strategies for mitigating the impact of pollution in air and water. Traditional sensors generally lack powerful processors and hence are only capable of capturing raw data from surroundings, which is further transmitted to remote stations with powerful architecture for final processing [4]. Such networks are subject to processing limitations, communication speed conflicts, and software compatibility constraints.

Significant innovations emerging through architecture are leading existing technology to intelligent sensor networks [5]. Intelligent design has resulted in smart, tiny, and low power electronic devices that are not only suitable for real-time sensing and monitoring purposes but also for control and processing-intensive tasks. This has unlocked several new domains of sensor network applications, including device networking, multilevel security (national defence services, office/home safety), optimized farming, factory automation and control, natural hazard prediction, social sensing, and other utilities, such as building automation, computerized meter readings, and intelligent automobiles (road traffic monitoring, congestion avoidance, and management).

Equipped with autonomous execution capability, smart sensors are empowered to process their own sampled data along with the collective coordination received from their immediate environment. Profusely scattered smart sensors (or sensor dust) allow enhancing the reliability and quality of service with minimum expenditure. The confluence of advanced nanotechnology and sensor electronics has heavily boosted the processing intelligence and input/output (I/O) capabilities of the sensor motes [6]. In the near future, intelligent sensor networks will provide an interface for remotely interacting with real-world entities and physical phenomena.

DOI: 10.1201/9781003427780-4

2.1.1 SMART HOMES

Sensor network technology in homes enables automation and enhances convenience for the owners. A smart home consists of a collection of internetworked sensing devices that are integrated into an automation system for imparting efficiency and security [7]. Figure 2.1 shows the basic layout of the electronic devices that capture data from the activity of the occupants in a smart home environment.

The sensor data gathered by these sensors is processed, analyzed, and further used to control and automate various functionalities of home electronics. Some widely used sensors for smart home and office automation include the following:

- Motion detection sensors: Motion sensors are designed to identify motion within a pre-determined region of interest. Such sensors are often used together with occupancy sensors to automate lighting, tracking activities of patients, movement of players in gaming consoles, and other security purposes [8]. These devices operate by discerning alterations in infrared radiation or other environmental elements triggered by the mobile entities in the vicinity. Common instances of motion sensors include passive infrared sensors, microwave sensors, and ultrasonic sensors. There are also hybrid technology sensors, which may embed microwave and passive infrared sensors for improved accuracy and are less prone to false alarms.
- Environment parameter sensors: There are certain sensors that can observe environmental parameters from the surrounding ambiance, such as temperature, air pressure, and humidity variations. This sensory information can be employed to automate heating, ventilation, and air conditioning

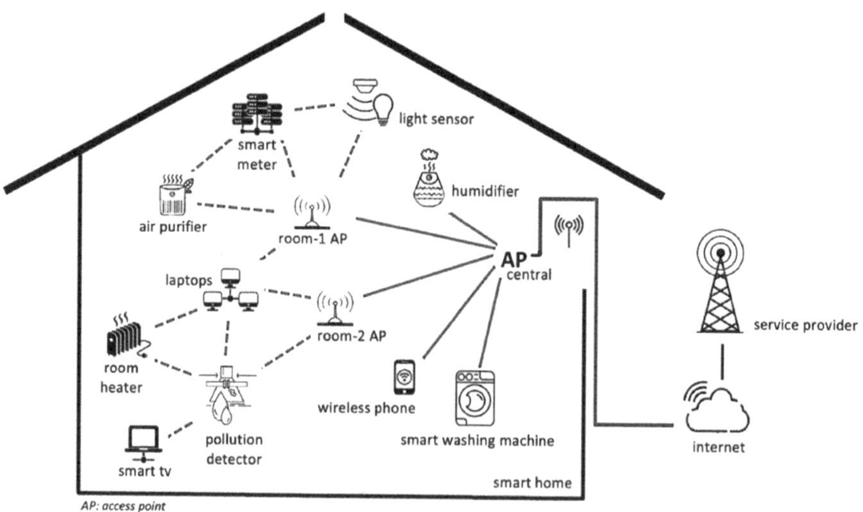

FIGURE 2.1 Smart home sensor network with electronic gadgets

operations in home and office environments. Such electronic equipment not only raises comfort levels but also leads to optimization of energy efficiency.

- Pollutant detectors: Recently, we have been confronting big challenges related to air pollution [9]. Hence, sensors are required for reporting early detection of smoke, carbon monoxide, and other atmospheric particulates. Such sensors may automate the operation of air purifiers to ensure the safety of the home occupants.
- Smart camera sensors: Camera sensors are crucial devices that can provide real-time monitoring of surroundings for the purposes of safety and concern of children, elderly people, and pets staying alone. These electronic devices are often coupled with sound sensors to provide rich audio-visual information and also generate voice commands [10]. This includes charge-coupled device sensors, complementary metal-oxide-semiconductor sensors, time-of-flight sensors, thermal imaging sensors, full-frame sensors, and infrared sensors. Such smart camera-enabled sensors can capture video footage and provide the data for further processing with computer vision and machine learning algorithms. Hence, this can produce advanced applications such as face recognition of visitors and monitoring their activities for enhanced security.
- Water leakage detection sensors: The water sensors installed in leakage-prone areas in smart homes can detect the presence of excess or unwanted water flows and can trigger alarms to prevent wastage of water [11]. An application of such devices may involve automatically switching off the supply when the water reservoirs become full.
- Smart electronic outlets: These sensor devices can be integrated with existing appliances, thereby allowing homeowners to remotely control and monitor the power usage of connected devices. This not only enables automation and convenience, but also power management and security against overuse of electrical devices.

Therefore, the smart home sensors, when integrated within a centralized hub, can allow setting of automation rules and the building of customized scenarios for monitoring home systems remotely through mobile or web applications. The goal is to impart intelligent automation, thus contributing to an overall improvement in the quality of life.

2.1.2 SMART HEALTHCARE

With regard to smart healthcare, sensor technology enables the collection and processing of real-time data, monitoring patients for improved diagnostics and thus enhancing overall healthcare management [12]. Sensor networks play an important role in remote patient monitoring and fall detection, thereby elevating the standards of patient care and safety. Sensor networking in the medical field is largely assists in monitoring, treatment, and prevention. There exist different types of pervasive monitoring support extended by smart electronics, which include:

- Remote monitoring.
- Continuous monitoring.
- Medication monitoring.
- Rehabilitation monitoring.
- Hygiene monitoring.

Remote monitoring of patients allows healthcare providers to remotely monitor body vital signs such as temperature, blood pressure, heartbeat rate, and sugar levels [13]. This vital health data is periodically collected and sent to medical practitioners and family caregivers to keep track of the health status of critical patients and chronic conditions of elderly people. This further assists in facilitating early detection of health-related abnormalities. Moreover, remote monitoring also helps in reducing the frequent need for visiting hospitals. This type of monitoring is often supported by wearable medical sensors for continuous monitoring of health metrics [14]. Such devices form an intelligent home remedial network to provide timely interventions through real-time analysis and tracking the health condition of patients. Additionally, there exists sensor-enabled devices to track pill dispensers so as to ensure the regularity and adherence of patients to the recommended medication regimens. Wireless sensor networks are also capable of monitoring rehabilitation environments by tracing the periodic advancements of individuals during therapy and diagnostic procedures [15]. This enables healthcare practitioners to change rehabilitation programs according to the reactions and advancements specific to patients. In addition to treatment-related monitoring facilities, there are smart sensing devices that can contribute to infection control by monitoring daily hygiene practices, which further helps to minimize the chances of infection spread, thus ensuring patient safety.

As depicted in Figure 2.2, sensor networks can be useful in several ways for providing smart healthcare delivery and executing critical treatment processes. This may include assistive devices in prosthetics, personalized treatment plans, and telemedicine support. Smart sensors can be integrated into prosthetics to enhance their operation, comfort, and responsiveness. For instance, smart sensors in prosthetic limbs and joint replacements can help assist in early movements of patients, thus improving mobility. Apart from this, analyzing sensor data also allows for the development of customized and evidence-based treatment plans that can be tailored specifically to the rate of improvement in patients. Sensor networks may also assist in imparting telemedicine support to elderly people by transmitting a steady stream of patient health data to doctors and medical practitioners. This further helps in providing virtual consultations, remote diagnostics, and improved chronic condition management, ultimately reducing the need for in-person visits.

In addition to smart monitoring and devising personalized treatment plans, sensor networks can also help in the early detection and prevention of diseases, especially during epidemic and pandemic situations [16]. The sensors can continuously track biomarkers and physiological parameters, thus resulting in the early detection of body malfunctions. Apart from the use of sensors in homes for remote monitoring, there could be a set of sensor networks within hospital infrastructure to track the location of medical assets and facilities. Additionally, there could be sensor-enabled

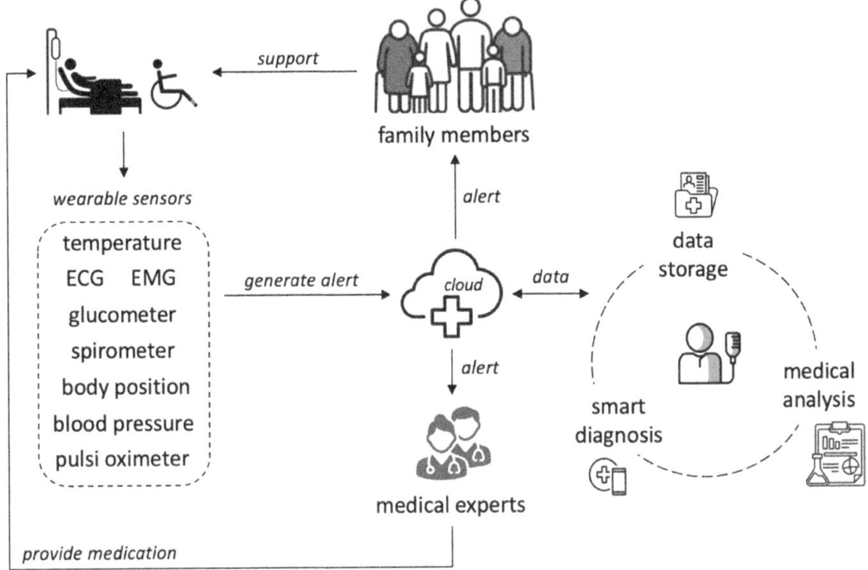

FIGURE 2.2 Smart healthcare for remote monitoring of patients

monitoring of medical equipment to further ensure reliability in healthcare operations. Within healthcare facilities and the residences of elderly individuals, sensor networks may incorporate accelerometers and motion sensors for fall detection. Swift identification of falls can prompt alerts to healthcare professionals or family members, ensuring timely assistance. Moreover, the use of biometric sensors, such as fingerprint scanners and iris recognizers, can be incorporated to guarantee secure access to patient data. Overall, in smart healthcare, the crucial role of sensor networks stems from their capacity to revolutionize conventional healthcare approaches, thus establishing a more patient-centred, data-driven, and efficient healthcare ecosystem. With the utilization of sensor technologies, healthcare providers can elevate patient outcomes, decrease healthcare costs, and enhance the overall quality of healthcare delivery.

2.1.3 INTELLIGENT FARM SENSING

Sophisticated farm sensing enabled through networks of sensors plays a vital role in precision agriculture [17]. The importance of sensor networks in this scenario is driven by their capacity to collect real-time observed data, monitor environmental conditions, and enhance different facets of farming activities. The sensor networks greatly facilitate precision agriculture with information about soil conditions, moisture levels, temperature, and other atmospheric variables. This data empowers farmers to fine-tune the utilization of resources such as water, fertilizers, and pesticides, resulting in heightened efficiency and diminished adverse environmental impact.

One of the significant aspects of precision agriculture involves monitoring crops, weather, livestock, and farming equipment. Smart farm sensors play an essential part in monitoring the health, growth, and production of crops [18]. This critical information is considered instrumental in the early detection of problems in crops, such as diseases, nutrient deficiencies, or pest infestations, allowing for timely intervention and thus minimizing potential crop losses. In addition to crop monitoring, sensor networks encompass real-time weather data concerning the farm location. This information holds significant value for scheduling agricultural activities, anticipating extreme weather events, and adjusting farming practices accordingly in response to changing weather conditions. Another application of sensor networks is livestock management, which involves overseeing the health and well-being of animals involved in farming. This encompasses tracking their movements, monitoring vital signs, and maintaining optimal conditions in barns or grazing areas. Farmers have the advantage of remotely overseeing and managing diverse aspects of their farm using mobile applications or web interfaces. This proves highly advantageous for large-scale operations, especially when handling multiple farm fields. Additionally, sensors can be deployed to monitor the functioning of farming equipment by tracking the usage frequency and patterns of different farm machinery. Such sensory information enables predictive maintenance, ultimately reducing downtime and preventing expensive equipment failures.

Apart from the monitoring tasks, the sensors networked over a large-scale farm area can also gather extensive data from the field and empower the farmers to make informed decisions based on real-time insights for smart harvesting, yield optimization, resource optimization, and pest control (Figure 2.3). For instance, utilizing soil moisture sensors enables the monitoring of moisture levels in the soil. This data assists farmers in optimizing irrigation schedules, ensuring that crops receive the appropriate amount of water at the right times. This practice is crucial for maintaining crop health and promoting water conservation. It also helps in performing intelligent pest control based on soil and crop conditions. By deploying sensors to identify pest activity, farmers can implement precise and timely interventions, thereby reducing the reliance on broad-spectrum pesticides [19]. This targeted approach enhances efficiency while minimizing the impact of pest-control measures.

Continuous monitoring of environmental conditions by sensor networks assists farmers in optimizing resource usage, ensuring efficient and sustainable practices [20]. This optimization extends to the judicious use of water, fertilizers, and energy, promoting environmentally friendly farming practices. Additionally, leveraging sensor data enables the creation of yield maps and the optimization of planting patterns. This empowers farmers to identify high and low-performing areas within the field, adapt planting strategies accordingly, and ultimately maximize the overall crop yield. In a nutshell, the importance of sensor networks in intelligent farm sensing

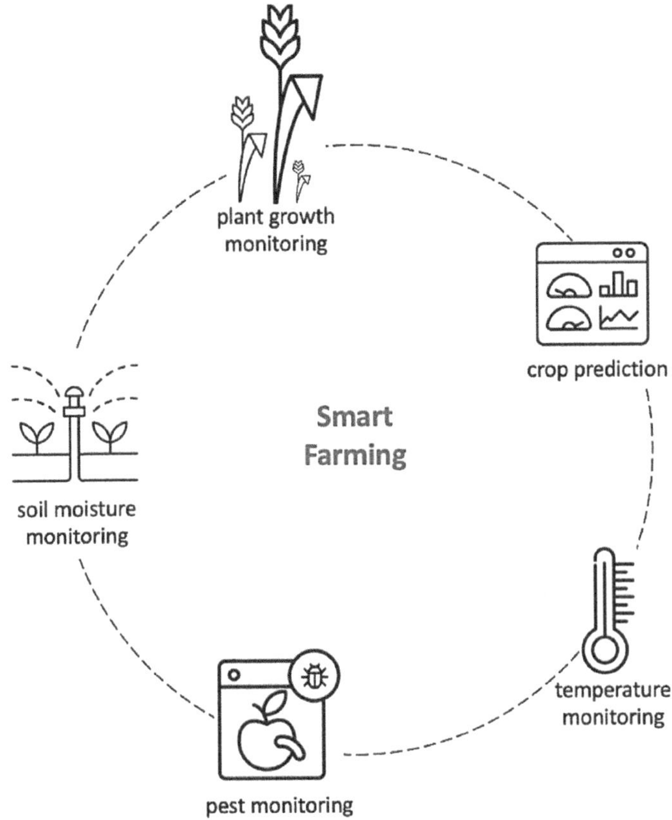

FIGURE 2.3   Smart farming techniques for precision agriculture

lies in their capability to revolutionize traditional farming practices, enabling data-driven and sustainable agriculture.

2.1.4 Terrestrial Sensor Network

Terrestrial sensor networks, encompassing a network of sensors deployed on the surface of Earth, hold significant importance across various applications. These include natural disaster detection, environmental monitoring, agriculture, wildlife monitoring, and forestry. In addition to this, smart city applications also form part and parcel of terrestrial networks. In urban areas, such wireless sensor networks are utilized for diverse purposes, such as traffic monitoring, geospatial mapping, air quality assessment, waste management, and tracking energy consumption [21]. This helps in advancing smart cities by furnishing sensory data to assist in efficient urban planning and city-wide resource management. Moreover, the sensor devices can be strategically installed on bridges, roads, and high-rise buildings to monitor the structural conditions, earthquakes and vibrations, stress levels, etc.

Such continuous sensory monitoring would ensure infrastructural safety before they escalate into serious problems. Also, sensors deployed across terrestrial landscapes play a pivotal role in the surveillance of public spaces, critical infrastructures, and national borders. These sensors are capable of detecting and alerting authorities to unauthorized activities or potential security threats, enhancing overall security measures in sustainable societies.

2.1.5 UNDERWATER SENSOR NETWORK

Underwater wireless sensor networks (UWSNs) are often deployed with the aim of exploring and monitoring aquatic activities underwater. Figure 2.4 shows the basic framework of a wireless sensor communication network with wireless floating nodes, anchored sensors, and sinks. These wireless sinks can be mobile or stationed at a position to support the monitoring of a particular underwater region. The aquatic sinks are further connected to terrestrial on-shore sinks and satellites for storage and sharing of the gathered data. With the aquatic range of sensing electronics and communication devices, these networks offer valuable data for scientific research, underwater monitoring, defence, and other undersea industrial applications [22].

- Aquatic explorations: The UWSNs enable research scientists to explore the vast and inaccessible underwater kingdom of aquatic flora and fauna. The acoustic sensors collect data on ocean currents, temperature, salinity, pressure, marine life, etc. This ultimately helps in a better understanding of oceanography and marine ecosystems (Figure 2.4). It can also further assist researchers in exploring sunken shipwrecks, ancient ruins, and other archaeological artefacts.
- Marine monitoring: The underwater network of sensors is often deployed for monitoring the health of marine environments [23]. The sensors can detect changes in water quality, temperature, acidity, pollution levels, and the impact of climate change on marine ecosystems. Further, the electronic devices can track fish movements for efficient aquaculture practices, such as

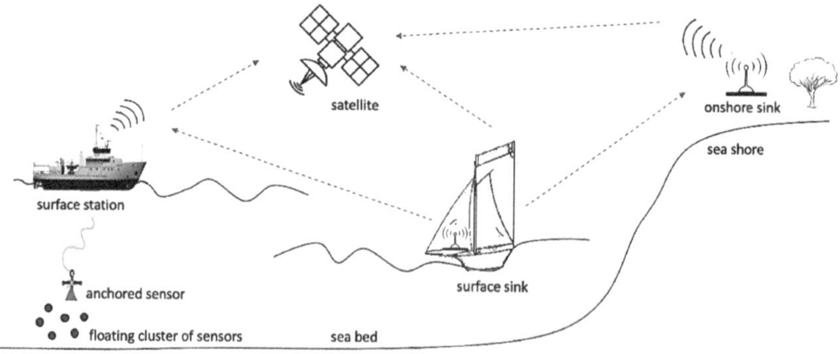

FIGURE 2.4   Underwater wireless sensor communication network

the optimization of feeding fish. Furthermore, UWSNs can assist in analyzing the impact of climate change on marine ecosystems and global climate patterns.

- Underwater surveillance and defence: The aquatic sensor network can also be designed and deployed for underwater surveillance and security applications, such as monitoring ports and the inspection of underwater infrastructures. They can generate alarm signals to alert the authorities about unauthorized activities, intrusions, and other underwater threats. Underwater sensor networks can also extend support for submarine detection, harbour safety, undersea mapping, and navigation. Overall, UWSNs contribute to underwater surveillance and situational awareness, especially for defence purposes.
- Undersea industrial sites: Apart from surveillance and monitoring, acoustic networks have immense applicability to underwater industries, such as oil and gas reserves, communication pipelines, and undersea drilling operations [24]. The sensors can be deployed for such underwater industrial applications to ensure the safety and integrity of subsea installations, oil leakage detection, the maintenance of cables for submarine communication, fault detection, etc. This further facilitates timely management and repair of critical aquatic operations having industrial importance.

2.1.6 Underground Sensor Network

There exists another category of WSN known as underground sensor networks, which holds immense value in observing and gathering information from below-ground areas, such as soil, underground tunnels, mineral mines, and other subterranean regions [25]. The significance of such underground wireless networks spans across multiple sectors, thus enhancing the management and safety of critical resources located in underground areas [26]. There are several areas and applications that use the power of sensors for underground asset management. This includes the archaeological study of burial sites with historical importance, monitoring soil quality, underground water supplies, oil pipelines, and sewage systems.

The use of underground sensors for the detection of leaks in pipelines can prevent the environment from becoming contaminated. Apart from environmental and commercial usage, underground sensors have immense significance in agriculture. The sensors deployed underground can measure soil type, moisture content, nutrient concentrations, and temperature, offering essential data for precision agriculture [27]. This information would further empower agriculturists to enhance irrigation, fertilization, and overall crop management, thereby resulting in increased crop production and resource efficiency.

2.1.7 Intelligent Vehicle Sensors for ITS

Intelligent vehicular sensor networks are pivotal in the broader domain of intelligent transportation systems (ITS). The use of vehicular sensors can elevate safety, efficiency, and the overall effectiveness required for smart transportation [28]. The importance of sensor networks becomes apparent in diverse applications within the domain of intelligent transportation. Such sensor devices are strategically integrated into vehicles to contribute to real-time data collection and analysis. The gathered data enables advancements in traffic management, collision avoidance, and transportation performance optimization. From vehicle detection and identification to adaptive-cruise control systems, vehicular sensors facilitate the pervasive and seamless flow of information critical for road safety. This further contributes to reducing traffic congestion on roads, especially at junctions and hotspot areas [29, 30]. The interconnected nature of these sensor networks forms the foundation of a highly responsive and efficient infrastructure for road, air, and water transportation. Apart from enhancing performance, the vehicular sensor network provides an interconnected transportation ecosystem that can be customized for the comfort and safety of travellers and pedestrians. The following are some important services offered by the intelligent vehicle sensors:

- Parking automation and assistance.
- Multi-point intersection management.
- Pedestrian detection and protection.
- Emergency vehicle service.
- Vehicle pathway monitoring.
- Fuel-efficient practices.

There exists a wide range of ultrasonic sensors and high-resolution visual sensors that can provide essential information about the surrounding proximity of the vehicles, thus assisting in automated parking. Moreover, there can be visible and audible alerts generated to warn the driver of potential collisions or obstacles during the parking process. In addition, vehicle-to-everything (V2X) technology, enabled by sensors and communication devices, allows automobiles to exchange information both between themselves and with base stations installed at intersections or blind turnings [31]. V2X can also facilitate quicker emergency response to ambulances, thus providing them a faster route. Additionally, as highlighted in Figure 2.5, intelligent sensors can be stationed on highways to help in the detection of pedestrians or animals suddenly crossing the vehicle path, and ultimately triggering automated braking systems in cars to prevent collisions. Road condition monitoring is another aspect of vehicular sensors that can detect environmental factors such as ice, snow, and wet surfaces, and alert drivers to activate safety measures such as anti-lock braking systems. This can foster enhanced traffic efficiency, decreased congestion, and heightened safety. Furthermore, the in-built sensor system can provide feedback on fuel efficiency to ensure reduced fuel consumption and emission-free driving practices.

2.1.8 Sensors for Automation and Control

Industrial processes have several highlighted instances that demand an indispensable requirement for sensor networks for process automation and control. Smart industrial

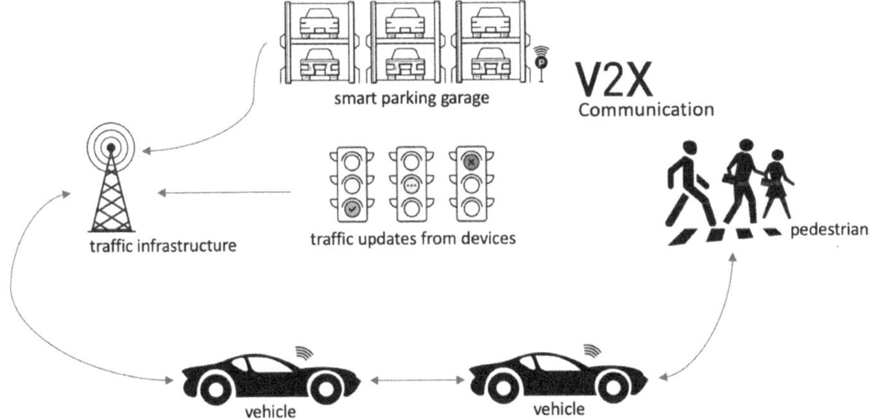

FIGURE 2.5 Vehicle-to-everything (V2X) communication network

sensors are challenged with critical monitoring of devices producing dynamic heterogeneous signals. Industrial sensors offer a high sampling rate to ensure accurate sensing and reliable reporting of continuously varying real-time phenomena [32]. Frequent sampling and data communication rapidly deplete sensor energy, which eventually impedes operational lifetime. Moreover, smart sensors are generally distributed in bulk to monitor industrial processes. Hence, low power design of hardware architecture and protocols needs to be considered for sensors. Energy efficiency is considered crucial to optimize the cost involved in sustaining sensor lifecycle, the absence of which the value of sensor networks would be lost in the context of industrial usage. In the manufacturing sector, market producers mainly focus on reaching targets and gaining profits while product servicing remains an overhead for customers. Once the products are bought, the management and maintenance of each piece of equipment become the burden of the customer. This adversely affects market policies and eventually leads to an imbalance between pricing and sales. To balance this, equipment networking is emerging as a viable solution. Advanced electronics implanted within each device allow the device to monitor its own functioning to achieve an extended operational lifetime with reduced power consumption. As illustrated in Figure 2.6, embedded intelligence also allows every unit to be networked with the manufacturing server, such that functional status, position control, location information, and surrounding measurements (operating temperature, humidity, etc.) can be reported via the cloud [33]. The equipment status reports and critical specifications would help producers to instigate research on the performance and reliability offered by their manufactured devices.

On the basis of received status updates, the need for repairing or replacing the product (or one of its parts) can be remotely identified and diagnosed accordingly. This can also improve reverse engineering to design and improvise equipment with higher quality service and customized benefits. In a nutshell, the origination of equipment networking has profitably attempted to bridge the gap often witnessed

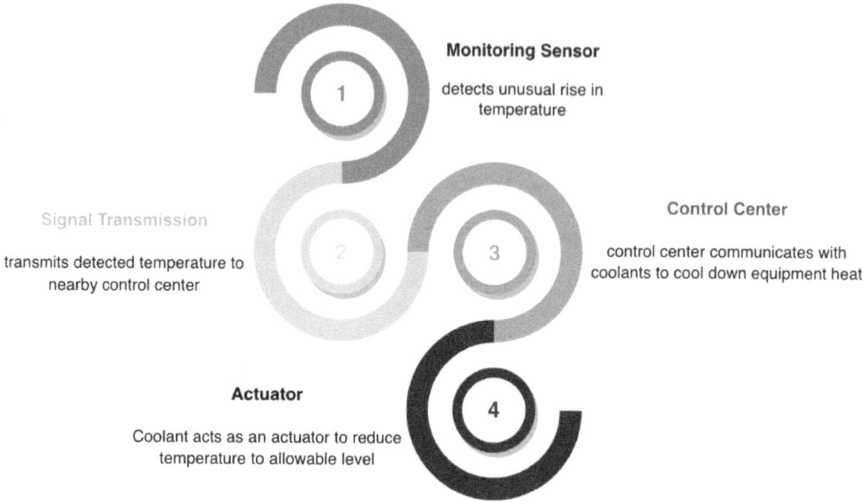

FIGURE 2.6 Equipment networking using sensors for automation and control

between product manufacturers and their clients. Inter-connected self-monitoring devices with installed intelligence eliminate maintenance hassles and significantly benefit manufacturers, suppliers, and consumers.

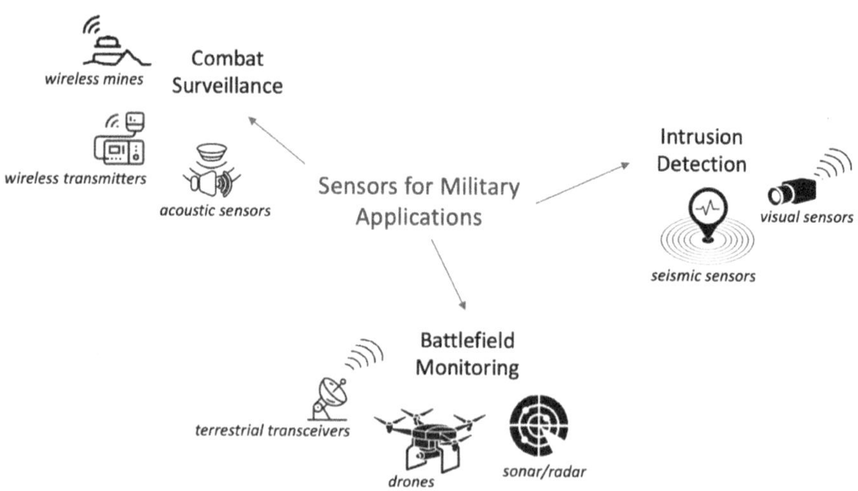

FIGURE 2.7 Sensor electronics for military operations

2.1.9 Sensors for Military Surveillance

In military surveillance, sensors facilitate detection, tracking, and analysis of diverse elements across the battlespace. Their importance in military surveillance is manifold, actively contributing to situational awareness, threat detection, effective coverage, and connected internetworking [34]. As seen in Figure 2.7, there are early warning systems that utilize sensors such as radar, sonar, drones, visual, seismic, thermal, and infrared. These devices are considered instrumental in identifying potential threats such as missiles, aircraft, unmanned aerial vehicles (UAVs), and unidentified submarines. Timely detection through these sensors enables military forces to promptly implement defensive measures against incoming threats, thus assisting in smart and robust decision-making. The sensors are capable of intercepting and analyzing electronic signals for monitoring communication signals, radar emissions, and electronic warfare activities to gather intelligence on enemy capabilities and intentions. Such smart electronics can also be used to detect and identify potential chemical, biological, radiological, and nuclear (CBRN) threats on the battlefield. This includes the use of chemical sensors, radiation detectors, and biological sensors to ensure the safety of military personnel. The amalgamation of data from diverse sensors and sources facilitates thorough analysis. Through data fusion, the accuracy and reliability of intelligence are heightened, presenting military decision-makers with a more comprehensive perspective for situational awareness of dynamic situations.

2.1.10 Hyperspectral Sensors for Remote Observation

The hyperspectral sensors are widely used in capturing data across numerous narrow and contiguous spectral bands, thereby enabling a meticulous analysis of the electromagnetic spectrum. This granular spectral information proves to be invaluable for characterizing materials, identifying substances, remote sensing, and gaining profound geologic insights into the composition of observed objects [35, 36]. These sophisticated imaging devices are capable of collecting data across an extensive range of wavelengths and offer detailed insights into the spectral properties of a scene. The importance of hyperspectral sensors in remote observation is also significant, given their unique capabilities that find application in various fields, including environmental monitoring, agriculture, and defence. In environmental monitoring, these types of sensors are employed for capturing detailed information concerning vegetation health, alterations in land cover, and water quality. Their particular utility lies in assessing ecological conditions, monitoring deforestation, and evaluating the aftermath of natural disasters. These electronic devices are also employed in ecological and biodiversity studies by providing detailed information about vegetation types, species diversity, and habitat conditions. They hold information focused on conservation and the ongoing monitoring of ecosystems.

The hyperspectral data allows observation of deforestation, land cover monitoring, and planning of land usage. This is considered significant for advancing sustainable land management methods, thereby strengthening conservation efforts. Also, hyperspectral sensors prove essential in disaster response and recovery. Such sensor devices are capable of evaluating damage due to natural disasters and identifying potential hazards, thus assisting in devising effective recovery strategies and enhancing military reconnaissance capabilities. Moreover, in defence applications, hyperspectral devices are often deployed for real-time critical problems, such as target identification, camouflage detection, and other anomalous activity identification.

2.1.11 Satellite Sensors for Space Exploration

Satellite sensors are pivotal in space exploration, offering vital data for scientific research of distant celestial objects, planetary observation, space monitoring, spacecraft navigation, exoplanet study, and extra-terrestrial exploration of life outside our solar system, etc. [37]. A wide range of satellite sensing devices are available to extensively cover various research facets, thus enhancing the ability to comprehend our contribution to the cosmos. These special sensors may include multispectral scanners, spectroradiometers, radiometers, and magnetometers. Planetary observing satellites are often built with various sensors that can provide valuable data for monitoring seasonal changes, climatic variations, weather patterns, natural disasters, etc. [38, 39]. Such sensory data can be further processed to generate insightful information for safeguarding our habitat from potential hazards. For instance, solar storms can be predicted by periodically monitoring solar activities, magnetic fields, cosmic radiation, solar irradiance, etc. Another useful application would be collision avoidance of satellites and spacecrafts by tracking space debris and objects in the Earth's orbit, thus ensuring the safety of operations in space. Apart from disaster management and hazard monitoring, satellite sensors are also used to provide navigation capabilities for spacecraft, rovers, and other space exploration vehicles that require precise trajectory calculations. Furthermore, satellites with specialized sensors can assist in space mining and resource exploration. This may involve identifying potential mineral deposits on celestial bodies in space.

2.2 INTEGRATION OF SOCIAL NETWORK WITH SENSORS

The increasing social relevance of data-intensive applications, such as environmental sensing, human/machine condition monitoring, and secured communication, has led to the revolutionary emergence of sensor networks [40]. This convergence of social networks and sensor technology results in the formation of a social network of sensors (SNoS) [41]. The socially networked wireless sensors help in the interpretation of locally sensed, context-aware data gathered by sensors that have ubiquitously infiltrated our everyday lives. The social network acts as a mining infrastructure for the bulk of contextual data produced by individual sensor nodes. This socially harvested information imparts agility to different societal applications with rich contextual awareness and improved personalization. This integrated technology helps to provide improved solutions for our personal benefit by utilizing the global understanding of local sensory data.

SNoS often models the sharing of information, diffusion of ideas, choice of careers, presence of chemicals, monitoring water levels, etc. Most of these applications rely on the structure of the underlying network to provide conclusive findings on market strategy, trading, employment, pollution, disaster awareness, etc. This necessitates exhaustive research on the modelling and analysis of the network structure that largely influences social behaviour.

2.2.1 Random Graphs Formation

The significance of modelling allows for the hypothetical creation of a simple representation of a complex network. It also enables the derivation of the model properties, which can be further used to predict outcomes under certain constraints [42]. An appropriate analysis of the proposed model can finally help to establish correctness in the representation of the real-world network.

The Erdo$\ddot{s}$ Renyi (E–R) random graph is one of the simplest ways to model a large complex social network of sensors (or individual entities). The model has the following assumptions:

a. Initially, there are N disconnected nodes in the network.
b. Eventually, an undirected network is formed with nodes connecting independently in a random manner.
c. Each node has (N–1) attempts to get edges.
d. Each attempt is considered a success with probability p.
e. The binomial distribution gives the probability that a node possesses degree d.

$$B_{N-1,p}(d) = \binom{N-1}{d} p^d (1-p)^{N-1-d} \tag{2.1}$$

The parameter p acts as a threshold that may be the outcome of a distribution function, or the flipping of an unbiased coin to decide the potentiality of an edge; i.e. with outcome $\geq p$, an edge is created, and vice versa. For a largely dense network, i.e. large N and small p, the above degree distribution in binomial formulation can be approximated with a Poisson distribution as

$$P_{N-1,p}(d) = \left[\frac{(N-1)^d}{d!}\right] \cdot p^d \cdot e^{-(N-1)p} \tag{2.2}$$

The random graph with degrees Poisson distributed is well known as the E-R Poisson random network or Bollob's $\ddot{a}$ network configuration, having an expected degree $p(N-1)$. In an E-R random graph, the largest component encompassing a significant portion of the network is referred to as the giant component (GC). An important property of GC is that it always occupies a finite fraction, even if the network tends to grow infinitely.

2.2.2 Degree Distribution in Random Graphs

The degree distribution exhibits the global pattern of the network; for instance, nodes (or people) who are well connected are referred to as hubs of information. Similarly, the shortest path between any two nodes, known as the geodesic, measures how evenly the network is distributed.

Theorem 1 (for densely connected network):

For a network with a large number of nodes N, it is assumed that links are dense enough such that the network is assumed to be connected, i.e. $d(n) \geq (1+\varepsilon)\log n$ for $\varepsilon > 0, \gg n$ where $d(n)$ is the degree of node $n \in N$; however, $\dfrac{d(n)}{n} \nrightarrow 0$ if the network is not fully connected, then the average path length and diameter are approximately proportional to $\dfrac{\log(N)}{\log(d)}$.

Proof A: In this case, we consider a restricted form of graph known as a Cayley tree in which each node, besides leaf nodes, has degree d (a binary tree is a special case of Cayley tree with degree, $d=2$). As per the property of Cayley trees, starting from the root, d nodes can be reached in level 1, $d(d-1)$ nodes can be reached in level 2, and so on. Therefore, moving out from the root in each direction, until l^{th} level, the total nodes that can be reached are: $d + d(d-1) + d(d-1)^2 + \ldots + d(d-1)^{l-1}$. This resembles a geometric progression with the first term d and ratio $(d-1)$, then the sum of the series becomes

$$N = \frac{d\left[(d-1)^l - 1\right]}{d-2} = \frac{d(d-1)^l}{d-2} - \frac{d}{d-2} \tag{2.3}$$

Usually, d is quite large, which means that $\left(\dfrac{d}{d-2}\right) \to 1$. Therefore, Equation (2.2) becomes:

$$N = \lim_{\substack{\left(\frac{d}{d-2}\right) \to 1 \\ d \gg l}} \frac{d(d-1)^l}{d-2} - \frac{d}{d-2} = (d-1)^l \Rightarrow \log N = l \log(d-1) \tag{2.4}$$

$$\therefore l = \frac{\log N}{\log(d-1)} \sim \frac{\log N}{\log d} \tag{2.5}$$

Proof B: In this proof, we deal with the case that the given graph is not necessarily in the form of a tree but an _E-R_ random graph in which node degrees are randomly generated. Our assumption is a large random graph, $d \geq (1+\varepsilon)\log N$, where d is the degree of a node that can be easily replaced by $E[d]$, the expected degree d of any node in the network. Now, it would be sufficient if it can be shown that the _fraction of nodes that possess nearly average degree goes to 1._

The above proof is stated using Chernoff bound. X is a binomial random variable (RV) which gives the interpretation of the expected number of links X is supposed to end up with. On applying the Chernoff bound, the probability that node X manages to obtain between three times and one-third of the expected number of links is at least $\left(1-e^{-E[X]}\right)$. Mathematically,

$$\Pr\left(\frac{E[X]}{3}\leq X\leq 3E[X]\right)\geq 1-e^{-E[X]} \tag{2.6}$$

Using the above formula, the probability of ith node to have a degree close to the average degree is given by

$$\Pr\left(\frac{d}{3}\leq d_i\leq 3d\right)\geq 1-e^{-d} \tag{2.7}$$

Similarly, the probability that all the nodes have their degrees lying approximately close to the average degree can be formulated as:

$$\Pr\left(\frac{d}{3}\leq d_{i|(\forall i\in N)}\leq 3d\right)\geq\left(1-e^{-d}\right)^N \tag{2.8}$$

In this case, an approximation is performed through probability because the degree is not independent but rather approximately independent. Plugging the given assumption, $d\geq(1+\varepsilon)\log N$, into above equation, we get:

$$\Pr\left(\frac{d}{3}\leq d_{i|(\forall i\in N)}\leq 3d\right)\geq\left(1-e^{-\log N^{1+\varepsilon}}\right)^N=\left(1-e^{\log\left(\frac{1}{N}\right)^{1+\varepsilon}}\right)^N=\left[1-\left(\frac{1}{N}\right)^{1+\varepsilon}\right]^N$$

$$\Rightarrow\Pr\left(\frac{d}{3}\leq d_{i|(\forall i\in N)}\leq 3d\right)\geq\left[1+\left(\frac{1}{N}\bullet\left(-N^{-\varepsilon}\right)\right)\right]^N \tag{2.9}$$

Using identities $\left(1+\frac{1}{N}\right)^N=e$ and $\left(1+\frac{x}{N}\right)^N=e^x$ with $x=-N^{-\varepsilon}$ in Equation (2.9),

we obtain the following expression in Equation (2.10).

$$\Pr\left(\frac{d}{3}\leq d_{i|(\forall i\in N)}\leq 3d\right)\geq e^{-N^{-\varepsilon}} \tag{2.10}$$

However, as the number of nodes increases $(N\to\infty)$, we have $e^{-N^{-\varepsilon}}\to 1$. This implies that the probability for degrees of all the nodes lying within a factor of 3 goes to 1, thereby signifying the correctness of the approximation. Now for large random networks, the expected degree can be approximated as $E[d]\sim N^{1/l}$ (where l is the level traversed outward from the root as specified in proof A). Therefore, our above formation evolves as in the following equations (2.11–2.14):

$$\frac{d}{3} \leq d_{i|(\forall i \in N)} \leq 3d \tag{2.11}$$

$$\Rightarrow \frac{d}{3} \leq E[d] \leq 3d \Rightarrow \frac{d}{3} < N^{1/l} < 3d \tag{2.12}$$

$$\Rightarrow \log\frac{d}{3}\left\langle \frac{1}{l}\log N \left\langle \log 3d \Rightarrow \frac{1}{\log\dfrac{d}{3}} \right\rangle \frac{l}{\log N} \right\rangle \frac{1}{\log 3d} \tag{2.13}$$

$$\Rightarrow \frac{\log N}{\log\dfrac{d}{3}} > l > \frac{\log N}{\log 3d} \tag{2.14}$$

Therefore, the approximate distance l taken by a node to get to any other node in the network, having the property that it lies within a factor of 3 with probability 1, can be bounded as in the above equation. Now, the degree of a node signifies the number of connections an entity in the social world keeps on gaining from time to time. So, for d, which is rapidly growing, both factors $\log\dfrac{d}{3}$ and $\log 3d$ become proportional to $\log d$, i.e. $\log\left(3d\, or\, \dfrac{d}{3}\right) = \log d \pm \log 3 \propto \log d$. Hence, even for the generalized case of $E\text{-}R$ random networks, we established the correctness of the theorem, $l \sim \dfrac{\log N}{\log d}$.

2.2.3 Isolation Probability in Random Graphs

For a large connected network modelled in the form of the $E\text{-}R$ graph, $E[d] \geq (1+\varepsilon)\log N$, it follows that $E[d] = \log N$ can be regarded as a threshold above which each of the nodes is expected to have some links. In other words, beyond this threshold, the possibility of having disconnected components eventually vanishes. Therefore, we have

$$E[d] = r + \log N \tag{2.15}$$

Also, it is known that for a densely connected $E\text{-}R$ random graph, we can formulate the average degree as

$$E[d] = p(N-1) \tag{2.16}$$

Now, the probability that some node is isolated is the same as the probability that it has no links. Now, p denotes the probability of the presence of any link; thus the probability of the absence of any link would be given by $(1-p)$. Since links are created independently of each other, the probability of a node i to be isolated (i.e. not connected) from the rest of the $N-1$ nodes in the network is given as in the following equation:

$$P_{iso}(i) = (1-p)^{N-1} = \left[1 - \left(\frac{r + \log N}{N-1}\right)\right]^{N-1} \tag{2.17}$$

The isolation probability $P_{iso}(i)$ approaches $e^{-r-\log N} = \dfrac{e^{-r}}{e^{\log N}} = \dfrac{e^{-r}}{N}$. Further, the expected number of isolated networks can be derived as

$$N_{iso} = \sum\nolimits_{i=1}^{N} P_{iso}(i) = \sum\nolimits_{i=1}^{N} \frac{e^{-r}}{N} = e^{-r} \tag{2.18}$$

It can be clearly interpreted from the above equation that, as $r = E[d] - \log N$ tends to infinity, the expected number of isolated nodes tends to zero, i.e. it provides an instance of an extremely well-connected network. However as $r \to -\infty$, average number of isolated nodes increases infinitely.

2.2.4 SENSORS AND NODE CENTRALITY

Centrality refers to the positional importance of nodes in a social network of sensors, which are more likely to influence several network functionalities, such as cluster coordination and closed community detection [43]. Centrality offers different measures to qualify certain nodes in the network for carrying out complex operations and communication tasks [44]. These central nodes are considered to be highly influential in the given network topology and are thought to perform more efficient sharing of information than peripheral nodes. The following section illustrates some of the significant approaches to capture the centrality measure of a node.

(a) *Degree centrality*: This refers to the topological measure that specifies the extent to which the local neighbourhood of a node is well-connected. It provides relevance to a node by estimating the size of its 1-hop neighbourhood. In other words, the node with the largest number of neighbours reachable in 1-hop is assigned the highest centrality. In order to scale the estimation of the centrality measure within 0 and 1, a normalization factor $n-1$ is considered, which signifies the maximum number of possible neighbours a node i may possess.

$$D_i(g) = \frac{d_i}{n-1} \tag{2.19}$$

where n refers to the total sensor nodes, and d_i is the degree associated with each node i in the network graph, $(g \rightarrow N,E)$, with N and E referring to the set of nodes and edges, respectively.

(b) *Closeness centrality*: This reveals the importance of a node in terms of its reachability from other nodes in the network. It assigns maximum significance to the node that requires fewer hops on average to reach any other node in the network. This measure can be formulated as

$$C_i(g) = \frac{n-1}{\sum_j l(i,j)} \tag{2.20}$$

where $l(i,j)$ denotes the shortest path distance between any node $j \in N$ to node i.

(c) *Decay centrality*: This helps to measure the decreasing influence of n-hop neighbours of a node as n increases. This presents an intuitive way of capturing the idea that nearest neighbours have higher probabilities of successfully delivering data with respect to a particular node i. However, as the distance increases, collision probabilities also tend to increase, thereby reducing the potential impact of further neighbours than those lying nearer to node i. Therefore, decay centrality $Dc_i^\delta(g)$ can be framed as the ratio of summation of the decaying impact of all the neighbours of node i over the normalization factor $(n-1)\delta$.

$$Dc_i^\delta(g) = \frac{\sum_{j \neq i} \delta^{l(i,j)}}{(n-1)\delta} \tag{2.21}$$

(d) *Betweenness centrality*: This designates a node to be highly central if it lies in the maximum number of shortest paths connecting any distinct pair of nodes in the network g. Mathematically, betweenness of a node h is expressed as the ratio $\dfrac{l_h(i,j)}{l(i,j)}$ for all pairs of nodes, normalized over $n-1$ nodes taken 2 at a time.

$$B_h(g) = \frac{\sum_{i,j \neq h} \dfrac{l_h(i,j)}{l(i,j)}}{\dbinom{n-1}{2}} \tag{2.22}$$

In the above equation, $l_h(i,j)$ refers to the number of shortest paths between nodes i and j that contain node h, and $l(i,j)$ denotes the total shortest paths existing between i and j.

(e) *Information centrality*: This assigns importance to a node i by measuring the drop in the network efficiency caused by removing i from the original network g. This is synonymous with the act of sleep scheduling some node i, such that the network efficiency is not compromised. This centrality can be formulated as

$$Ic_i(g') = \frac{\Delta N_{eff}}{N_{eff}(g)} = \frac{N_{eff}(g) - N_{eff}(g')}{N_{eff}(g)} \tag{2.23}$$

where $N_{eff}(g)$ is the efficiency of the original network g, and $N_{eff}(g')$ refers to the modified graph $g' \to (N-i, E-\{e_i\})$, which excludes node i and all the edges incident on it. Further, the network efficiency $N_{eff}(g)$ can be stated as

$$N_{eff}(g) = \frac{1}{n(n-1)} \sum_{\substack{i \neq j \\ (i,j \in N)}} \varepsilon_{ij} \tag{2.24}$$

where ε_{ij} is the efficiency parameter i and is defined as the inverse of the shortest path from all its directly as well as indirectly connected neighbour(s) j, i.e. $\varepsilon_{ij} = \dfrac{1}{l(i,j)}$.

2.2.5 SOCIAL SENSING APPLICATIONS

The proliferation of numerous social networking portals has led to a tremendous increase in communication between people across the world by exchanging information, opinions, likes, comments, etc. [45]. People, nowadays, are more active on their virtual social sites rather than in their actual social life. Initially, communication of ideas was accomplished in the form of textual mode by writing text, which was later revolutionized by the invention of emoticons and smileys for emphasizing human expressions. Social sensing involves the study of detecting the human emotions, sentiments, and perceptions by analyzing text, posts, transcripts, etc. Apart from plain text, there is increasing usage of phrases, symbols, and emoticons, as different forms of expressing emotion and human sentiments in text. Social sensing has gained increasing significance in contexts such as social media monitoring, perceptions of newly launched products, feedback analysis of customers, disaster response analysis, corporate team formation, chatbot reply enhancements (especially for healthcare and wellness), and personalized content recommendations.

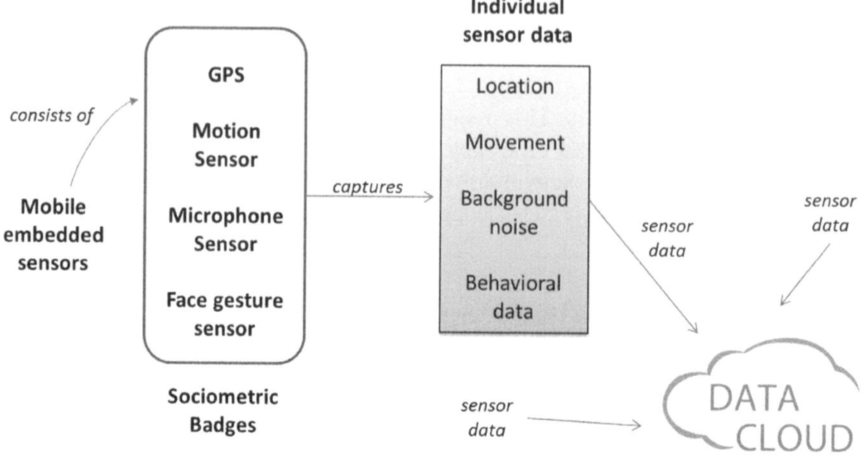

FIGURE 2.8   Social sensing with sociometric devices

The growth of this upcoming technology is fuelled by embedding location aware-ness, mobility detection, voice/face recognition, etc., over portable mobile devices known as *sociometric badges* [46]. As shown in Figure 2.8, the features captured by these sensing units can be further used for detecting geographic position, move-ment, voice pitch, social behaviour, and sentiment, etc. This complete set of raw data drawn from an individual bearing the mobile device can be uploaded to the cloud. Moreover, the sharing of information can be controlled by the user depending on what he wants to dedicate to the cloud. As the amount of data outsourcing from the mass increases, powerful processors can be deployed to generate aggregated results and inferences. Though the high-end cloud processing of data remains opaque to users, the inferences drawn from the giant processors help people to make important decisions, for instance, avoiding lanes with heavy traffic movement, visiting highly rated restaurants and shopping outlets, choosing parks and sanctuaries located away from city noise, etc.

Medical science can also benefit from using the intelligence of social sensing. Patients wearing sociometric badges can be monitored by detecting any abnormal biomedical activities or by comparing sentimental analysis statistics of regular days with the symptomatic ones. A patient with challenging health conditions might often isolate himself from the crowd, preferring less physical movement [47]. Generally, comparative voice data can also be useful in signalling deteriorating body condi-tions by monitoring pitch, frequency, and duration of conversations with others. This would help provide optimized and timely diagnosis and health relief to the patients (Figure 2.9).

Another important application of social sensing can be observed in companies where a number of engineers are grouped together to collaborate on some projects demanding quality outcomes with tough deadlines [48]. In such cases, sociometric data together with one's own performance and social data (obtained from social

FIGURE 2.9 Application domains of social sensing

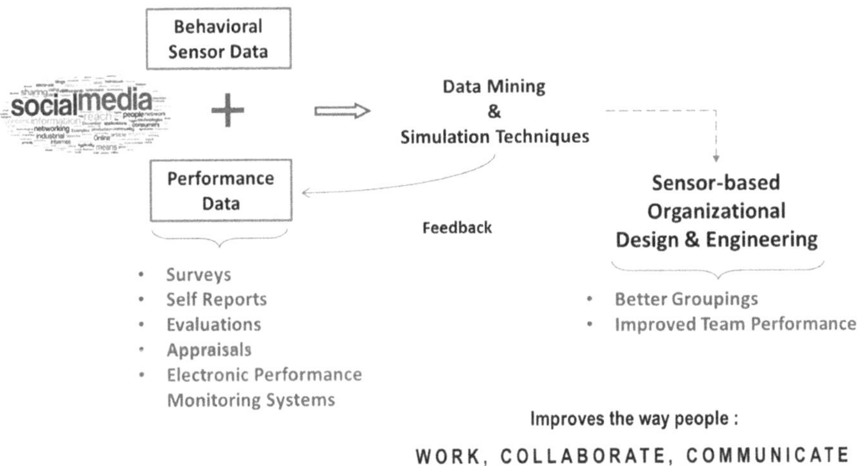

FIGURE 2.10 Organization performance analysis with social sensing

media, such as Facebook and Twitter) can be used to identify a set of individuals as a team that can help achieve improved organizational performance (Figure 2.10). Therefore, social sensing would redefine personal interactions with corporate benefits by enhancing the way people communicate, collaborate, and work.

2.3 INTEGRATION OF NEURAL COMPUTING WITH SENSORS

The integration of neural computing with sensors involves combining neural computing models with sensor technologies to enhance the capabilities of sensing systems. This integration can further improve decision-making in various applications.

Artificial neural networks can be employed to process and analyze the raw sensory data collected from the environment and identify complex patterns or anomalies in sensor data [49] and, therefore, to extract relevant features from sensor inputs. This is particularly valuable in applications such as adaptive learning, where the operating conditions are subject to frequent changes in the incoming sensor data. In such cases, the power of dynamic sensor input to neural networks can help computational systems evolve and improve over time. When working with real-time data, many other services can also be performed with sensor-driven neural computing, such as anomaly detection in smart industries for predictive maintenance of equipment [50]. Such monitoring also minimizes downtime and reduces maintenance costs of the potential equipment. Moreover, sensors can be deployed in power grids to record electricity consumption, energy storage, power management, grid monitoring, and security. Neural networks can further analyze this data to optimize energy distribution, thus leading to efficient and reliable smart grids. Neural networks can also be optimized for efficient processing, enabling energy-efficient sensor systems. This is particularly important in applications where power consumption is a critical factor, such as in wireless sensor networks or IoT devices. It can also facilitate real-time decision-making based on sensor inputs for applications such as smart healthcare and home automation, where quick responses to changing conditions are essential.

Another widely used concept is neuromorphic computing, which is inspired by the structure and operations of the human brain. It involves designing hardware and algorithms to simulate neural networks [51]. Integrating neuromorphic computing with sensors can lead to brain-like efficiency. Moreover, brain-computer interfaces incorporated with neuro-sensor electronics can enable direct communication between the brain and external devices. There are certain sensors available to capture brain signals using electroencephalogram electrodes. The data collected from such interfaces can be utilized to empower neural network systems for further interpreting the captured signals. Some significant applications in this context may include prosthetic limb control, virtual reality, and paralysis assistance. Moreover, in wearable devices, sensors such as accelerometers and heart rate monitors can collect data about activities and physiological parameters of an individual. This gathered sensory input can be used by neural networks to detect patterns indicative of health conditions and provide personalized health recommendations on hygiene, diet, exercises, and other essential routines.

The usefulness of integrating neural computing with sensors shows immense potential for improved adaptability in highly dynamic environments. For instance, neuro-sensing is extensively found in smart living environments. In human-computer interaction, sensors such as cameras or depth sensors capture gestures and movements. Neural networks can be trained to recognize and interpret such gestures, thus enabling hands-free control of devices. This can also give impetus to virtual reality and other forms of natural human interactions. Sensors deployed in smart homes, including motion sensors, temperature sensors, and cameras, capture data about the environment and occupants. It is clear that sensor networks are highly instrumental in the transformation of diverse areas of human activities, thereby fostering a globally connected and data-driven landscape.

REFERENCES

1. Borges, Luis M., Fernando J. Velez, and Antonio S. Lebres. "Survey on the Characterization and Classification of Wireless Sensor Network Applications." *IEEE Communications Surveys & Tutorials* 16, no. 4 (2014): 1860–90.
2. Martinez, K., J. K. Hart, R. Ong, S. Brennan, A. Mielke, D. Torney, A. Maccabe, M. Maroti, G. Simon, A. Ledeczi, and J. Sztipanovitz. "Sensor Network Applications." *IEEE Computer* 37, no. 8 (2004): 50–6.
3. Nassra, Ihab, and Juan V. Capella. "Data Compression Techniques in IoT-enabled Wireless Body Sensor Networks: A Systematic Literature Review and Research Trends for QoS Improvement." *Internet of Things* (2023): 100806. https://doi.org/10.1016/j.iot.2023.100806.
4. Singhal, Shikha, Adwitiya Sinha, and Buddha Singh. "Context Awareness for Healthcare Service Delivery with Intelligent Sensors." *Frontiers of Data and Knowledge Management for Convergence of ICT, Healthcare, and Telecommunication Services* (2022): 61–83. https://doi.org/10.1007/978-3-030-77558-2_4.
5. Anand, L., S. Padmalal, J. Seetha, R. Juliana, PS Naveen Kumar, and Gayatri Parasa. "Evaluation of Wireless Sensor Networks Module Using IoT Approach." In 2023 Third International Conference on Artificial Intelligence and Smart Energy (ICAIS), pp. 1543–6, IEEE, 2023.
6. Mahdavi Dehkharghani, Farnaz, Mobin Ghahremanlou, Zahra Zandi, Mahmoud Jalili, M. R. Mozafari, and Parisa Mardani. "Future Energy and Therapeutic Perspectives of Green Nano-Technology: Recent Advances and Challenges." *Nano Micro Biosystems* 2, no. 1 (2023): 11–21.
7. Malik, Indu, Arpit Bhardwaj, Harshit Bhardwaj, and Aditi Sakalle. "IoT-Enabled Smart Homes: Architecture, Challenges, and Issues." *Revolutionizing Industrial Automation Through the Convergence of Artificial Intelligence and the Internet of Things* (2023): 160–76. https://doi.org/10.4018/978-1-6684-4991-2.ch008.
8. Guan, Min, Yang Liu, Hong Du, Yinying Long, Xingye An, Hongbin Liu, and Bowen Cheng. "Durable, Breathable, Sweat-Resistant, and Degradable Flexible Sensors for Human Motion Detection." *Chemical Engineering Journal* 462 (2023): 142151.
9. Singh, Ragini, Akhela Umapathi, Gaurang Patel, Chayan Patra, Uzma Malik, Suresh K. Bhargava, and Hemant Kumar Daima. "Nanozyme-Based Pollutant Sensing and Environmental Treatment: Trends, Challenges, and Perspectives." *Science of the Total Environment* 854 (2023): 158771.
10. Stähler, Julian, Carsten Markgraf, Mathias Pechinger, and David W. Gao. "High-Performance Perception: A Camera-Based Approach for Smart Autonomous Electric Vehicles in Smart Cities." *IEEE Electrification Magazine* 11, no. 2 (2023): 44–51.
11. Sitaropoulos, Konstantinos, Salvatore Salamone, and Lina Sela. "Frequency-Based Leak Signature Investigation Using Acoustic Sensors in Urban Water Distribution Networks." *Advanced Engineering Informatics* 55 (2023): 101905.
12. Wilson, Marian, Roschelle Fritz, Myles Finlay, and Diane J. Cook. "Piloting Smart Home Sensors to Detect Overnight Respiratory and Withdrawal Symptoms in Adults Prescribed Opioids." *Pain Management Nursing* 24, no. 1 (2023): 4–11.
13. Shaik, Thanveer, Xiaohui Tao, Niall Higgins, Lin Li, Raj Gururajan, Xujuan Zhou, and U. Rajendra Acharya. "Remote Patient Monitoring Using Artificial Intelligence: Current State, Applications, and Challenges." *Wiley Interdisciplinary Reviews: Data Mining and Knowledge Discovery* 13, no. 2 (2023): e1485.
14. Al-Daraghmeh, Mohammad Y., and Richard T. Stone. "A Review of Medical Wearables: Materials, Power Sources, Sensors, and Manufacturing Aspects of Human Wearable Technologies." *Journal of Medical Engineering & Technology* 47, no. 1 (2023): 67–81.

15. Pellegrini, Michael, Natasha A. Lannin, Richelle Mychasiuk, Marnie Graco, Sharon Flora Kramer, and Melita J. Giummarra. "Measuring Sleep Quality in the Hospital Environment with Wearable and Non-Wearable Devices in Adults with Stroke Undergoing Inpatient Rehabilitation." *International Journal of Environmental Research and Public Health* 20, no. 5 (2023): 3984.

16. Pilato, Giovanni, Fabio Persia, Mouzhi Ge, Theodoros Chondrogiannis, and Daniela D'Auria. "A Modular Social Sensing System for Personalized Orienteering in the COVID-19 Era." *ACM Transactions on Management Information Systems* 14, no. 4 (2023): 1–26.

17. Sudharson, K., Badi Alekhya, G. Abinaya, C. Rohini, S. Arthi, and D. Dhinakaran. "Efficient Soil Condition Monitoring with IoT Enabled Intelligent Farming Solution." In 2023 IEEE International Students' Conference on Electrical, Electronics and Computer Science (SCEECS), pp. 1–6, IEEE, 2023.

18. Neethirajan, Suresh. "Artificial Intelligence and Sensor Innovations: Enhancing Livestock Welfare with a Human-Centric Approach." *Human-Centric Intelligent Systems* (2023): 1–16. https://doi.org/10.1007/s44230-023-00050-2.

19. Rajak, Prem, Abhratanu Ganguly, Satadal Adhikary, and Suchandra Bhattacharya. "Internet of Things and Smart Sensors in Agriculture: Scopes and Challenges." *Journal of Agriculture and Food Research* 14 (2023): 100776.

20. Ataei Kachouei, Matin, Ajeet Kaushik, and Md Azahar Ali. "Internet of Things-Enabled Food and Plant Sensors to Empower Sustainability." *Advanced Intelligent Systems* (2023): 2300321. https://doi.org/10.1002/aisy.202300321.

21. Wang, Dawei, Menghan Wu, Chinmay Chakraborty, Lingtong Min, Yixin He, and Manisha Guduri. "Covert Communications in Air-Ground Integrated Urban Sensing Networks Enhanced by Federated Learning." *IEEE Sensors Journal* (2023). https://doi.org/10.1109/jsen.2023.3322784.

22. Singhal, Shikha, Shairal Tanwar, and Adwitiya Sinha. "Ant Colony Optimization based Routing for Underwater Sensor Network." In 2020 9th International Conference System Modeling and Advancement in Research Trends (SMART), pp. 33–8, IEEE, 2020.

23. Gola, Kamal Kumar, and Shikha Arya. "Underwater Acoustic Sensor Networks: Taxonomy on Applications, Architectures, Localization methods, Deployment Techniques, Routing Techniques, and Threats: A Systematic Review." *Concurrency and Computation: Practice and Experience* 35, no. 23 (2023): e7815.

24. Saini, Preeti, Rishi P. Singh, and Adwitiya Sinha. "Survey of Radio Propagation Models for Acoustic Applications in Underwater Wireless Sensor Network." *International Journal of Sensors Wireless Communications and Control* 11, no. 1 (2021): 42–53.

25. Akyildiz, Ian F., and Erich P. Stuntebeck. "Wireless Underground Sensor Networks: Research Challenges." *Ad Hoc Networks* 4, no. 6 (2006): 669–86.

26. Silva, Agnelo R., and Mehmet C. Vuran. "Empirical Evaluation of Wireless Underground-to-Underground Communication in Wireless Underground Sensor Networks." In *International Conference on Distributed Computing in Sensor Systems*, pp. 231–44. Berlin, Heidelberg: Springer Berlin Heidelberg, 2009.

27. Tan, Xin, Zhi Sun, and Ian F. Akyildiz. "Wireless Underground Sensor Networks: MI-Based Communication Systems for Underground Applications." *IEEE Antennas and Propagation Magazine* 57, no. 4 (2015): 74–87.

28. Wang, Zhangu, Jun Zhan, Chunguang Duan, Xin Guan, Pingping Lu, and Kai Yang. "A Review of Vehicle Detection Techniques for Intelligent Vehicles." *IEEE Transactions on Neural Networks and Learning Systems* (2022). https://doi.org/10.1109/tnnls.2021.3128968.

29. Bishop, Richard. "A Survey of Intelligent Vehicle Applications Worldwide." In Proceedings of the IEEE Intelligent Vehicles Symposium 2000 (Cat. No. 00TH8511), pp. 25–30, IEEE, 2000.
30. Parent, Michel, and Patrice Bodu. "Intelligent Vehicle Technologies." In *Automotive Embedded Systems Handbook*, pp. 1–3. CRC Press, 2017.
31. Noor-A-Rahim, Md, Zilong Liu, Haeyoung Lee, Mohammad Omar Khyam, Jianhua He, Dirk Pesch, Klaus Moessner, Walid Saad, and H. Vincent Poor. "6G for Vehicle-to-Everything (V2X) Communications: Enabling Technologies, Challenges, and Opportunities." *Proceedings of the IEEE* 110, no. 6 (2022): 712–34.
32. Åkerberg, Johan, Mikael Gidlund, and Mats Björkman. "Future Research Challenges in Wireless Sensor and Actuator Networks Targeting Industrial Automation." In 2011 9th IEEE International Conference on Industrial Informatics, pp. 410–5, IEEE, 2011.
33. Reininger, T., F. Welker, and M. Von Zeppelin. "Sensors in Position Control Applications for Industrial Automation." *Sensors and Actuators A: Physical* 129, no. 1–2 (2006): 270–4.
34. Thomas, Diya, Rajan Shankaran, Mehmet Orgun, Michael Hitchens, and Wei Ni. "Energy-Efficient Military Surveillance: Coverage Meets Connectivity." *IEEE Sensors Journal* 19, no. 10 (2019): 3902–11.
35. Van der Meer, Freek D., Harald MA Van der Werff, Frank JA Van Ruitenbeek, Chris A. Hecker, Wim H. Bakker, Marleen F. Noomen, Mark Van Der Meijde, E. John M. Carranza, J. Boudewijn De Smeth, and Tsehaie Woldai. "Multi-and Hyperspectral Geologic Remote Sensing: A Review." *International Journal of Applied Earth Observation and Geoinformation* 14, no. 1 (2012): 112–28.
36. Bishop, Charlotte, Benoit Rivard, Carlos de Souza Filho, and Freek Van Der Meer. "Geological Remote Sensing." *International Journal of Applied Earth Observation and Geoinformation* 64 (2018): 267–74.
37. Brown, Christopher W., Laurence N. Connor, John L. Lillibridge, Nicholas R. Nalli, and Richard V. Legeckis. "An Introduction to Satellite Sensors, Observations and Techniques." In *Remote Sensing of Coastal Aquatic Environments: Technologies, Techniques and Applications*, pp. 21–50. Dordrecht: Springer Netherlands, 2005.
38. Schilling, Klaus. "Perspectives for Miniaturized, Distributed, Networked Cooperating Systems for Space Exploration." *Robotics and Autonomous Systems* 90 (2017): 118–24.
39. Katz, Daniel S., and Raphael R. Some. "NASA Advances Robotic Space Exploration." *Computer* 36, no. 1 (2003): 52–61.
40. Aggarwal, Charu C., and Tarek Abdelzaher. "Integrating Sensors and Social Networks." *Social Network Data Analytics* (2011): 379–412. https://doi.org/10.1007/978-1-4419-8462-3_14.
41. Rosi, Alberto, Marco Mamei, Franco Zambonelli, Simon Dobson, Graeme Stevenson, and Juan Ye. "Social Sensors and Pervasive Services: Approaches and Perspectives." In 2011 IEEE International Conference on Pervasive Computing and Communications Workshops (PERCOM Workshops), pp. 525–30, IEEE, 2011.
42. Kim, Ji Youn, Michael Howard, Emily Cox Pahnke, and Warren Boeker. "Understanding Network Formation in Strategy Research: Exponential Random Graph Models." *Strategic Management Journal* 37, no. 1 (2016): 22–44.
43. Sinha, Adwitiya, and D. K. Lobiyal. "Performance Analysis of Data Aggregation for Overlapping Clusters in Heterogeneous Sensor Network." In 2013 Sixth International Conference on Contemporary Computing (IC3), pp. 143–8, IEEE, 2013.
44. Rodrigues, Francisco Aparecido. "Network Centrality: An Introduction." *A Mathematical Modeling Approach from Nonlinear Dynamics to Complex Systems* (2019): 177–96. https://doi.org/10.1007/978-3-319-78512-7_10.

45. Zhang, Daniel, Yue Ma, Yang Zhang, Suwen Lin, X. Sharon Hu, and Dong Wang. "A Real-Time and Non-Cooperative Task Allocation Framework for Social Sensing Applications in Edge Computing Systems." In *2018 IEEE Real-Time and Embedded Technology and Applications Symposium (RTAS)*, pp. 316–26, IEEE, 2018.

46. Kawamoto, Eiji, Asami Ito-Masui, Ryo Esumi, Mami Ito, Noriko Mizutani, Tomoyo Hayashi, Hiroshi Imai, and Motomu Shimaoka. "Social Network Analysis of Intensive Care Unit Health Care Professionals Measured by Wearable Sociometric Badges: Longitudinal Observational Study." *Journal of Medical Internet Research* 22, no. 12 (2020): e23184.

47. Sharma, Pallavi, and Pankaj Deep Kaur. "Effectiveness of Web-Based Social Sensing in Health Information Dissemination—A Review." *Telematics and Informatics* 34, no. 1 (2017): 194–219.

48. Schmid Mast, Marianne, Daniel Gatica-Perez, Denise Frauendorfer, Laurent Nguyen, and Tanzeem Choudhury. "Social Sensing for Psychology: Automated Interpersonal Behavior Assessment." *Current Directions in Psychological Science* 24, no. 2 (2015): 154–60.

49. Zhou, Feichi, and Yang Chai. "Near-Sensor and In-Sensor Computing." *Nature Electronics* 3, no. 11 (2020): 664–71.

50. Rangwala, S., and D. Dornfeld. "Sensor Integration Using Neural Networks for Intelligent Tool Condition Monitoring." *Journal of Engineering for Industry* 112, no. 3 (1990): 219–28.

51. Davies, Mike, Andreas Wild, Garrick Orchard, Yulia Sandamirskaya, Gabriel A. Fonseca Guerra, Prasad Joshi, Philipp Plank, and Sumedh R. Risbud. "Advancing Neuromorphic Computing with Loihi: A Survey of Results and Outlook." *Proceedings of the IEEE* 109, no. 5 (2021): 911–34.

Chapter 3

SUMMARY

In the field of wireless sensor networks (WSNs), there are two broad categories: target coverage and area coverage. In this chapter, we address the target coverage problem where a set of targets are deployed in the pre-defined region along with the sensor nodes. The objective of these sensing devices is to monitor the targets for the longest possible duration, referred to as network lifetime. The sensors are equipped with a limited energy source (battery life) which is non-rechargeable and sometimes non-replaceable depending upon the network application. A target is said to covered as long as it is covered by at least one sensor. Hence, the scarce source of the network (battery) is the prime concern, as it has direct impact on the functional duration of the deployed network. There are many variants of target coverage, such as target full coverage, target K-coverage, target partial coverage, and target connected coverage. There are many algorithmic paradigms to solve all these variants so that network lifetime can be prolonged. Genetic algorithm-based metaheuristics are proven to be best-suited for solving the target coverage problem, and hence we have discussed some of these heuristics and analyze their comparative performance on deployed networks of different dimensions.

DOI: 10.1201/9781003427780-5

3 Coverage in Wireless Sensor Networks

3.1 INTRODUCTION: WIRELESS SENSOR NETWORKS

Nowadays, wired networks have almost been replaced by wireless networks due to current requirements. Today's examples of wireless communication are mobile phones, laptops, and cellular telephones. In general, wireless sensor networks (WSNs) have numerous components that work as sensor nodes. A sensor node generally communicates with neighbouring nodes to share or receive data to/from the central master node, which is generally known as the base station (BS). These sensors are embedded with a small inbuilt internal memory where users can store relevant information, a microprocessor that operates on collected data, and a small onboard battery that enables the sensor to work until the battery exhausts [1]. Unlike traditional wireless networks, sensors in WSNs are designed to communicate information to the central station, situated within the same proximity via 1-hop or multi-hop architecture. These sensing nodes do not talk directly to other nodes in the same network unless they are adjacent. Furthermore, the network is self-adaptable to compensate for the failure of any candidate node in the network [2].

There are diverse applications of sensor networks where WSNs are spread to monitor designated areas for the longest possible period, allowing each and every activity within that area to be observed. Among these applications of wireless sensor networks, a few others are indoor monitoring, military or sensitive security services, inventory tracking, wildlife and battlefield surveillance, and smart buildings [3]. Most of these applications require the active participation of the deployed nodes for the longest possible time so that the network remains functional to perform the desired task in a certain predefined manner. To be in functional mode, these sensing devices make use of the inbuilt battery resource, which makes them active during the desired period. The most important and critical component of a sensor node is the power source. As discussed earlier, a very small battery unit is attached to these sensor nodes, which makes them prone to failure which in turn renders the deployed network non-functional. There are certain applications of WSNs where the network is deployed in human-inaccessible terrain, making it impossible to replace the exhausted sensors' battery. To manage this issue, one must ensure that all sensing devices are used in optimized way to maximize their usage [4–6]. In order to utilize sensors in an energy-efficient manner, sensors have various states and can switch between states as necessary. Broadly, these sensor states are sleep, idle, transmit, and receive [7]. To enhance the deployed network lifetime (i.e. functional duration), one

DOI: 10.1201/9781003427780-6

has to optimize the usage of these nodes by switching their states one after another based on functionality. Another way of saving network power is by optimizing the route to the base station, enabling the least required intermediate sensors to forward sensed data from one place to another within the same network [8]. Further, network lifetime can be enhanced by activating only the necessary sensors at a given time. This way, one can certainly prolong the deployed network lifetime.

3.2 COVERAGE IN WIRELESS SENSOR NETWORKS

Deployed WSNs face major issues in measuring how well the designated objects are being monitored while the network is functional [9]. The sensors can cover an object that falls within their sensing range. An object is said to be covered by a sensor if its Euclidean distance is less than the range of the designated sensor. In general, the standard coverage concern in the field of wireless sensor networks is known as the target and area coverage problem.

Area coverage is a well-discussed and widely studied coverage problem in WSNs [10]. The major objective behind this is to simply monitor the designated region for the longest possible duration. To do so, sensors are densely deployed in that area so that the complete geographical region is covered by these sensors, as depicted in Figure 3.1. By observing this figure, we can say that not even a single corner of the predefined area is left uncovered by the deployed networks. However, there are many sensors that are covering redundant areas, which happens due to the dense deployment.

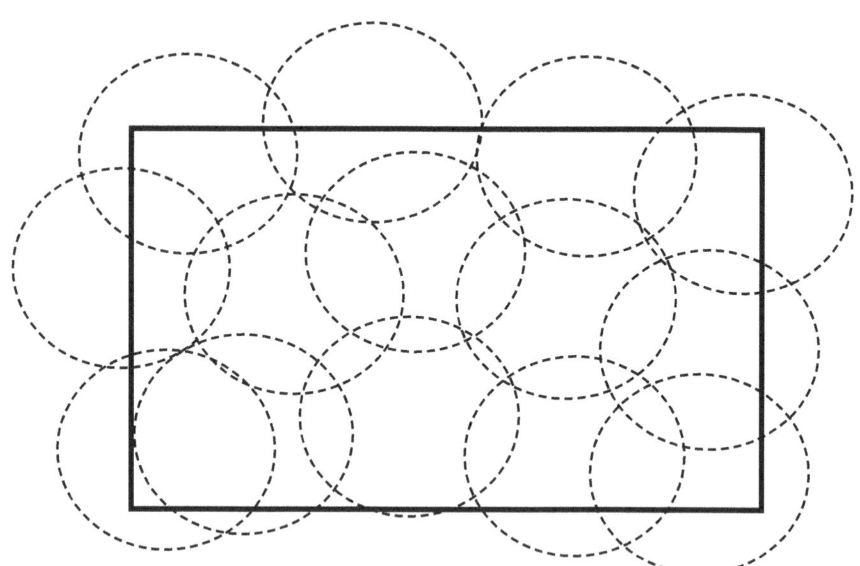

FIGURE 3.1   Area coverage

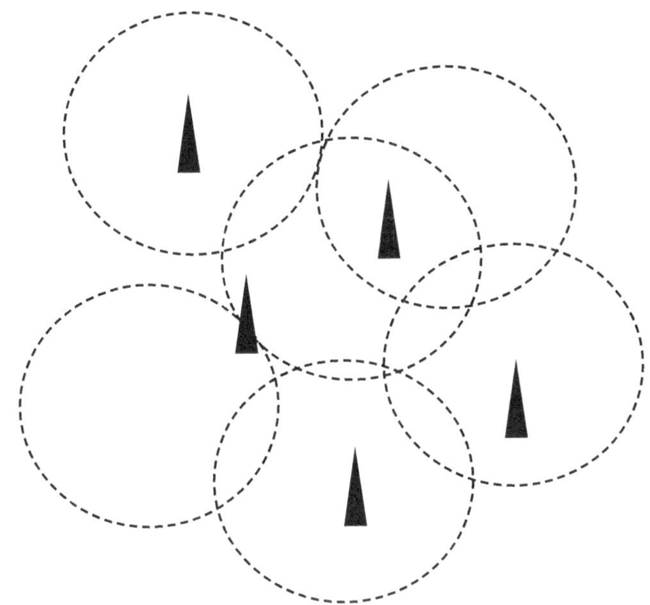

FIGURE 3.2 Target coverage

*Target **coverage*** in WSNs is basically aimed at monitoring a group of targets that are spread over the same region where the sensor network is deployed [11]. Figure 3.2 shows that all the targets (black nodes) are completely in the vicinity of sensors and all are covered by these sensors. Here, we can easily observe that none of the targets (out of five) are left uncovered by the deployed sensor nodes. Again, we can observe that many targets are redundantly covered in the densely deployed sensor networks. Such types of deployment are especially required at the moment because there are many applications where deployment is not an easy task, especially in wildlife terrains or dense forest areas.

These two area and target coverage problems ensure the designated area or targets are continuously covered. Only providing coverage is not sufficient when it comes to monitoring such geographical terrain where humans cannot reach to replace exhausted sensor batteries. Since sensor networks are energy constrained, we need to utilize their batteries in an optimized way to prolong network operational time.

3.3 ENERGY OPTIMIZATION IN COVERAGE

In a power-scarce sensor network, optimizing battery usage is the main concern, with many energy-efficient methodologies discussed by researchers in the past. To extend the lifetime of the deployed WSNs, certain scheduling techniques must be used to alternate sensors between active and sleep states. While a set of sensors is active, sensors with redundant coverage may enter a sleep state so that these sensors can save their battery for future coverage. To optimize battery usage, instead of activating all the sensors to cover the whole set of targets, a small subset of sensors is activated to cover the targets. These subsets are called sensor covers (or cover sets) with associated fixed active times. Once these possible cover sets are generated, their activation durations are summed to calculate network lifetime. Thus, the main objective behind such scheduling is to prolong the network lifetime by activating only a single cover set at a time. By doing so, one can devise energy-efficient coverage paradigms to facilitate the monitoring of the designated area or targets. Based on how these cover sets are designed, energy-efficient algorithms in the literature are further categorized as disjoint or non-disjoint methodologies and centralized or decentralized [12] methodology. In the case of disjoint methodologies, none of the sensors are part of multiple cover sets, whereas sensors can participate in multiple cover sets in non-disjoint paradigms to optimize their battery usage. The centralized paradigms ensure that all the deployed sensor nodes send data to the central unit, which we call the base station, while in the case of decentralized methodologies, there is no central unit where sensors send data for further processing; instead, data processing takes place in a decentralized manner. In most energy-optimized practices, centralized paradigms are preferred.

Among these two coverage problems, the target coverage problem is widely discussed, as most of the time, no one is interested in monitoring a huge geographical area, but users try to cover targets dispersed in the proximity. Now, we will discuss basic terms and definitions used throughout the chapter to understand the rest of the sections where the target coverage problem will be discussed in detail with the help of many energy-efficient paradigms for prolonging network lifetime. Before going into an in-depth discussion, first we present the target coverage problem in the field of wireless sensor networks and then we will discuss some required notations and definitions to understand it.

3.3.1 TARGET COVERAGE PROBLEM

Let $s1$, $s2$, ..., and sx be X sensor nodes randomly deployed in a given predefined area, which is rectangular in size, to monitor Y number of targets $t1$, $t2$, ..., and ty. All these sensing devices are equipped with battery life Ei. In general, it is assumed that sensor si covers a target tj if the target lies within the sensing range ri of that sensor. Furthermore, a cover set is known to be the sensor subset out of the total number of sensors (X) which are altogether responsible for covering all the targets (Y). In general, a cover set $SC = \{si \mid$ for each tj there is a $si \in SC$ such that si covers $tj\}$. The working duration of a cover set SC, $w(SC)$ should not exceed $Minsi \in SC\, Ei$.

3.3.2 DEFINITIONS AND NOTATIONS

Now, we will discuss some basic required definitions and notations to understand the terms used for solving the target coverage problem for WSNs, which are part of the subsequent sections of the chapter.

Definition 1: Sensor cover matrix (CR)
Let's define a matrix CR as

$$CR_{ij} = \begin{cases} 1, \textit{if sensors}_i \textit{is in sensor cover } SC_j \\ 0, \textit{otherwise} \end{cases} \tag{3.1}$$

The linear programming formulation of the same problem, which is stated above, is given below.

$$Maximize \sum_r wr \tag{3.2}$$

$$subject\ to \sum_r CR_{ir}x_r \le E_i \text{for all sensors } s_i \tag{3.3}$$

$$w_r \ge 0, \text{for all sensor covers } SC_r.$$

Here, we can notice that the objective of the above-stated linear program is mainly to enhance the total network lifetime of the given WSNs (i.e. the sum of all the cover sets' functional duration).

Definition 2: Coverage matrix (CM)

The coverage matrix, *CM*, represents the sensor target coverage status in the deployed network. Let $CM\left[X,Y\right]$, coverage matrix *CM* with X rows, number of sensors and Y columns, and the number of targets with all the entries either 0 or 1 as given below.

$$CM = \begin{cases} 1, \textit{if} \text{ sensor } s_i \text{covers target } t_j \\ 0, \textit{otherwise} \end{cases} \tag{3.4}$$

For example, consider the coverage matrix shown in Table 3.1, where there are four sensors and three targets. Here, 0 at the index (i, j) of *CM* shows that the sensor *si* does not cover the target, *tj*, and 1 means it covers it.

Definition 3: Sensor cover/cover set (SC)

This is a subset of sensors out of the total number of sensors (X) that jointly cover all the targets (Y). Thus, we can represent a cover set $SC = \{si|$ for each *tj* there is a $si \in SC$ such that *si* covers *tj* $\}$. To illustrate this formation, let us consider the network of four sensors and four targets as depicted in Table 3.2. Here, we can consider

TABLE 3.1

Coverage Matrix (CM) Representation

s/t	t_1	t_2	t_3
s_1	0	0	1
s_2	1	1	0
s3	1	0	0
s_4	0	1	0

TABLE 3.2

Cover Set Calculation

s/t	t_1	t_2	t_3	t_4
s_1	1	1	0	0
s_2	0	0	0	1
s_3	0	1	0	1
s_4	1	0	1	0

one cover set, $SC = \{s1, s2\}$. Since all the targets are covered by this cover set, we can find another cover set $SC = \{s3, s4\}$ which has different sensors. We can also observe here that both cover sets are disjointed in nature where none of the sensors are participating in more than one cover set.

Definition 4: Sensor cover lifetime $x(SC)$

The functional duration can be set for the generated cover equal to remaining smallest battery unit of the sensor that is participating in the given cover set. Formally, it can be represented as $x(SC)$, and cannot exceed $Min\{si \in SC\,Ei\}$. To explain it in detail, let us consider the following network, which includes five sensors to cover four targets as shown in Table 3.3. Here, besides the coverage, battery life (i.e. E) corresponding to each sensor is also mention in the same network. Furthermore, the given network is heterogeneous, where sensors are equipped with different battery lives. To understand this, let us consider one cover set $SC = \{s1, s2, s3\}$ where three sensors are capable enough to cover all these four targets. While calculating the cover set lifetime, $x(SC)$, we have to assign the smallest sensors' battery unit to it, which is part of this cover set. Thus, the cover set lifetime $(x(SC))$ over here is 1 unit, which is the lowest of these three participating sensors in the cover set as given below.

$$x(SC) = Min\{si \in SC\,Ei\} \tag{3.5}$$

In most of the target coverage scenarios, this method is used to calculate the total function duration of a given cover set.

Definition 5: Network lifetime (L)

This is the functional duration until the first target becomes uncovered as per the coverage variant requirement. It can be calculated by aggregating all the cover sets' lifetimes. To illustrate this concept, let us consider the given network of six sensors and four targets as shown in Table 3.4, where all the sensors are equipped with different battery units.

TABLE 3.3

Sensor Cover Life Calculation

E	s/t	t_1	t_2	t_3	t_4
$E_1=1$	s_1	1	1	0	0
$E_2=3$	s_2	0	0	1	0
$E_3=2$	s_3	0	1	0	1
$E_4=1$	s_4	1	0	1	1
$E_5=2$	s_5	1	1	0	0

TABLE 3.4

Network Life Calculation

E	s/t	t_1	t_2	t_3	t_4
$E_1=2$	s_1	1	1	0	0
$E_2=2$	s_2	0	0	1	0
$E_3=2$	s_3	0	1	0	1
$E_4=3$	s_4	1	0	1	1
$E_5=3$	s_5	0	1	0	0
$E_6=3$	s_6	1	1	0	0

Illustration 1

Here, let us say we can generate two cover sets, namely $SC-1$ and $SC-2$ as $SC-1=\{s1,s2,s3\}$ and $SC-2=\{s4,s5,s6\}$. The lifetime of $SC-1$ is 2 units, and lifetime of $SC-2$ is 3 units. So, the total network lifetime is 5 units over here.

Definition 6: Critical target ($T_{critical}$)

During coverage, in a homogeneous network the least covered target is known as a critical target, whereas for a heterogeneously deployed network, the critical target is the one whose coverage is lowest (sum of all the batteries of covering sensors). To illustrate this definition, let us consider two different networks where one is homogeneous, as shown in Table 3.5, and another is heterogeneous, as shown in Table 3.6.

The homogeneous network shown in Table 3.5 consists of four sensors and three targets, where every sensor is equipped with a unit battery life. Here, target $t3$ is a critical one, as it is covered by only one sensor (i.e. $s4$).

Similarly, we consider another network which is heterogeneous in nature where there are four sensors with different battery lives and three targets, as shown in Table 3.6. Here, target t_2 is critical because the summation of the battery lives of these sensors which are covering it is minimal compared other targets.

Thus, in general, the index of the critical target can be represented as follows.

$$j = \mathbf{argmin} \sum iCMijEi \tag{3.6}$$

TABLE 3.5

Critical Target in Homogeneous Network

s/t	t_1	t_2	t_3
s_1	1	1	0
s_2	0	0	0
s_3	0	1	0
s_4	1	0	1

TABLE 3.6

Critical Target in Heterogeneous Network

E	s/t	t_1	t_2	t_3
$E_1=1$	s_1	1	1	0
$E_2=3$	s_2	0	0	1
$E_3=2$	s_3	0	1	0
$E_4=3$	s_4	1	0	1

TABLE 3.7

Critical Sensor

s/t	t_1	t_2	t_3
s_1	1	1	0
s_3	0	1	0
s_3	1	0	1

Definition 7: Critical sensor ($S_{critical}$)

As discussed in *definition 6*, there is a concept of a critical sensor (s) in the deployed network. A critical sensor (s) is the sensor that covers a critical target (s). To illustrate this, let us consider the network scenario shown in the Table 3.7. As shown in this network, s_3 is covering the critical target $t3$, thus, it is a critical sensor.

Definition 8: Upper bound (UB)

In a wireless sensor network, the upper bound is basically decided by the coverage of the critical target. Thus, one can calculate the upper bound for the underlying networks' lifetime by considering the critical target(s) and their associated coverage.

Illustration 2

To understand the upper bound, we assume a network of five sensors and four targets which are deployed in the same vicinity, as shown in Table 3.8. All the sensors

TABLE 3.8
Upper Bound Calculation

E	s/t	t_1	t_2	t_3	t_4
$E_1=1$	s_1	1	1	0	0
$E_2=3$	s_2	1	0	1	1
$E_3=2$	s_3	0	1	1	1
$E_4=3$	s_4	1	1	0	1
$E_5=1$	s_5	1	1	1	1

are assigned initial battery life in a heterogeneous manner where sensors have different batteries. To calculate the working time of a cover set, we consider the method given in **(3.5)** where the functional duration of the cover set is calculated.

Here, in this network given as in Table 3.8, the coverage of each target is clearly seen. We can easily observe here that for this scenario, the critical target is t_3 whose coverage is 6 units. Hence, UB (the optimal upper bound) for the underlying network scenario is 6 units, and no methodology can achieve more than this. So, this is how one can calculate the network optimal upper bound and design their coverage heuristics in order to achieve their performance near the optimal upper bound. Formally, we can also denote the calculated upper bound as follows:

$$UB = Minj \sum i\, CMij * Ei \tag{3.7}$$

3.4   VARIANTS OF COVERAGE

Based on the type of coverage, there are mainly four coverage variants in WSNs. Broadly, we classify them as full coverage, *K*-coverage, partial coverage, and connected coverage. Now, one can discuss them in detail with reference to the target coverage problem.

3.4.1   ENERGY-EFFICIENT FULL COVERAGE

This refers to the scenario of a sensor network where a fixed number of objects (or targets) are scattered in the same field where sensors are deployed. Here, the major concern is monitoring all the deployed target nodes in the same network for as long as possible, as shown in Figure 3.3. Here, there are five sensors, namely *s*1,...,*s*5, and three targets, namely *t*1, *t*2, and *t*3. We can clearly observe from Figure 3.3 that all the three targets are covered. Further, the aim of the full coverage problem is to prolong the total network lifetime with the help of subsets of sensors (sensor cover/ cover set), where each standalone cover set is sufficient to cover all the objects [13]. We can also observe in Figure 3.3 that there are excessive sensor nodes compared

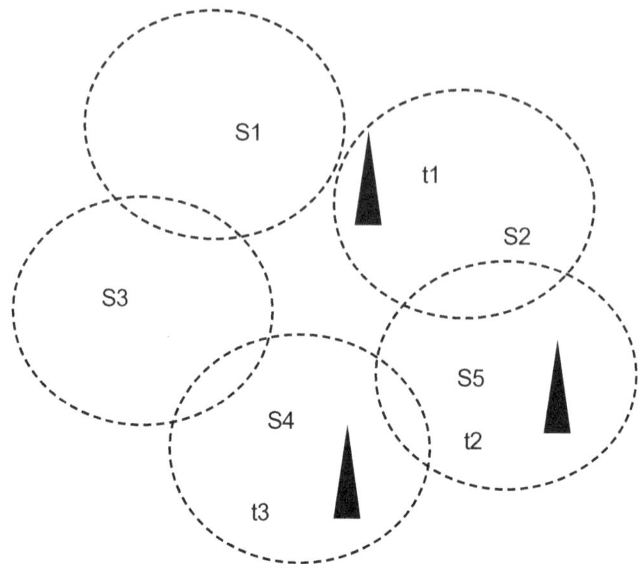

FIGURE 3.3 Full coverage

to the number of targets. The reason behind such a scenario is that the network is densely deployed to meet the desired requirements of an application.

There have been many studies done so far to provide energy-efficient target full coverage aiming to extend the total achieved network lifetime [12, 13, 19, 20]. All the studies done differ only by the way their protocols generate cover sets. The authors in [12] addressed the coverage problem and designed an energy-efficient algorithm where sensors with maximum coverage are given the highest priority while considering them as part of the cover set. The users repeatedly selected sensors with the highest remaining target coverage until a cover set was formed. By following this pattern, the deployed network will have energy holes at an early stage, and due to this, the optimal upper bound of network lifetime cannot be achieved. We have already discussed that as the deployed network has non-replaceable batteries, we have to maximize network lifetime. Thus, the addressed approach is not well suited for energy-scarce networks. To overcome this issue of energy holes, the authors in [13] discussed another energy-efficient heuristic which differs from the work in [12] the way they generated cover sets. The authors in [13] proposed a methodology for selecting cover sets where sensors with the highest residual energy are preferred when selecting them for the current cover set. By the time a cover set is generated, there is the possibility that some of the selected sensors are providing redundant coverage for targets. Since the network is energy-deprived all the time, the proposed methodology [13] further removes redundant sensors from the generated cover set to enhance the sensor coverage, which increases the achievable network lifetime. The authors in [19] designed another energy-efficient heuristic based on weight assignment to the sensors. This study also considered remaining energy as the prime criterion for selecting sensors. With this, if the theoretical upper bound (*UB*) is not achieved, then sensors are reassigned weights based on their coverage of targets, and same process is repeated until the achieved total lifetime is less than UB. Similarly, another research work is presented by authors in [20] where an energy-efficient approach is addressed to further enhance functional duration.

3.4.2 ENERGY-EFFICIENT K-COVERAGE

This refers to the class of coverage where the target is considered covered if and only if *K* distinct sensors are covering it in each individual cover set. Some crucial applications need this kind of coverage for better quality [14, 15]. For example, if we have to deploy CCTVs to provide continuous surveillance for a security room, then one has to ensure that each and every corner of the vicinity is monitored without fail for every fraction of a second. In such applications, we should monitor these hot spots with more than one sensor to ensure quality of service (QoS). As shown in Figure 3.4, we have considered four sensors, *s1,...,s4*, and three targets, namely *t1*, *t2*, and *t3*, and $K = 2$. It is clearly depicted in Figure 3.4 that there are at least two sensors covering each target (because here we choose $K = 2$). Depending on the application requirements, the value of K can vary within any range. For example, consider a security-critical application where redundant sensors cover many areas in order to ensure that all spots are covered all of the time. In such critical requirements, K can take any value as defined by the customers.

There have been many studies done so far with respect to *K*-coverage where the objective is mainly to maximize the total achievable network functional duration [21–23]. Pujari et al. [21] addressed this variant of coverage by proposing an

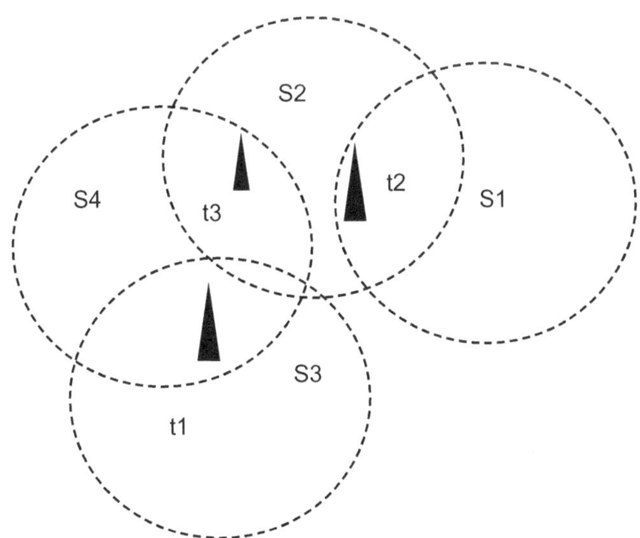

FIGURE 3.4 *K*-coverage

energy-based methodology to prolong it by the selection of sensors with the highest residual energy to form cover sets. After generating cover sets, they further removed extra sensors with redundant coverage from the cover set so that more sensors would be available for the consecutive cover sets. Thus, by selecting only the highest remaining energy sensors without redundancy, the authors achieved a lifetime closer to UB. Furthermore, the authors in [22] explored another method of maximizing total functional duration, which considers highest residual energy as well as sensors covering a maximum number of uncovered targets, giving priority to forming cover sets. By doing so, the authors ensured that their heuristic chose the minimum of number of sensors to form the cover, which further reduces the need to minimize the generated cover set, thereby speeding up the process in comparison to [21]. Hence, the proposed methodology has a good impact in prolonging the network lifetime while requiring less time required to achieve almost the same outcome as in [21]. Yu et al. [23] introduced an altogether new methodology to achieve a network lifetime near the optimal upper bound, where objective is to provide K-coverage. To form cover sets, the authors considered three major parameters to ensure K-coverage: the coverage of the sensors (i.e. how many uncovered targets are covered by a particular sensor), residual battery life, and K-coverage. Therefore, only those sensor nodes that fulfil all these three major parameters decided by the heuristic are selected as part of the cover set. Apart from proving K-coverage, the same study further ensured connectivity in the network.

3.4.3 Energy-Efficient Partial Coverage

This is another class of coverage where a few targets are not to be considered covered (i.e. a fraction of targets) in each sensor cover. It is known as α-coverage ($\alpha \in (0,1]$) where each sensor cover is should cover only $T\alpha$ targets ($T\alpha \subseteq T$). Here, T represents the actual set of targets in the network [16]. We generally need this type of coverage variant where even a few objects can be left uncovered, as it is not going to impact the cumulative performance of the application. For example, suppose we want to measure soil humidity over a period for weather forecasting. In such a scenario, we know that nearby sensors that are covering objects are providing almost the same data (or reading). Hence, as shown in Figure 3.5, we can afford to avoid a few objects, as their coverage hardly impacts the performance. As depicted in Figure 3.5, there are four sensors and three targets that are deployed in close proximity. Here, we see that target $t1$ is not covered by any of the sensors, which is altogether acceptable as per the predefined requirements of the application.

To understand this partial coverage variant's utility, consider Figure 3.6 where four targets and four sensors are deployed in the same vicinity. As compared to Figure 3.5, here we see all the targets are covered, but while generating cover sets, we will avoid a pre-defined number of targets left unattended in every sensor cover. We consider every sensing node to have unit battery life, and the functional duration of each cover set is considered as 1 unit while generating the cover set. Hence, the network lifetime achieved is at most 2 units in total with two cover sets. Now, if we

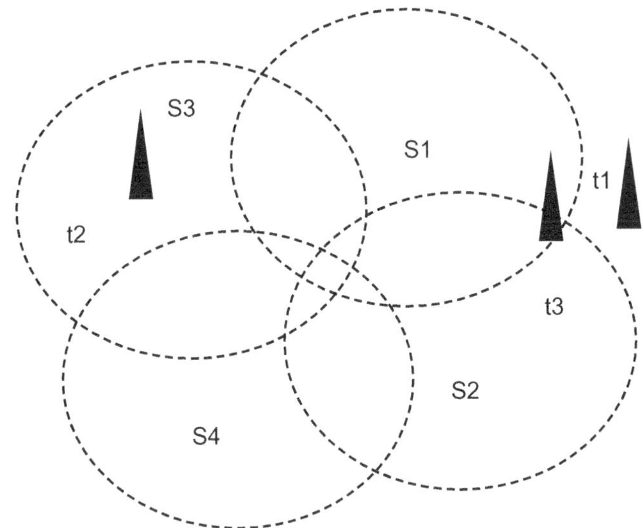

FIGURE 3.5 Partial coverage

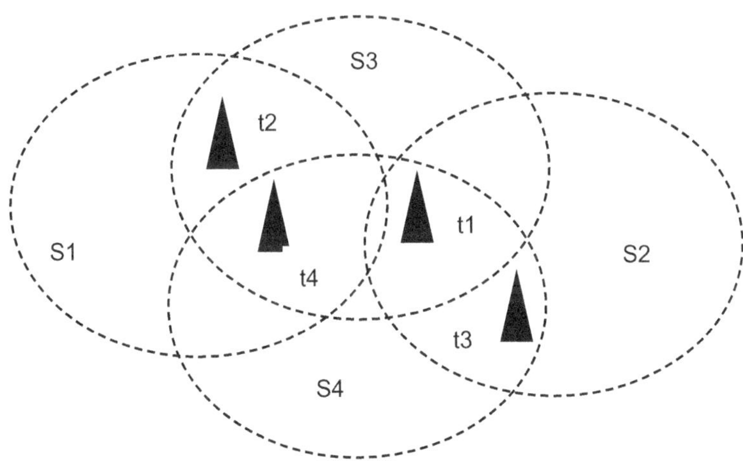

FIGURE 3.6 Partial coverage illustration

consider partial coverage with $\alpha = 0.75$, we can avoid one target out of these four in each cover, and that target can be left uncovered. The values of α can be pre-defined with respect to the application requirements. No rule exists to decide upon the values of α. To assign lifetime to each generated cover set, we follow the rule given in Equation (3.5) to remove ambiguity. With this partial consideration, three cover sets can be formed, namely $SC-1 = \{s1, s2\}$, $SC-2 = \{s3\}$, and $SC-3 = \{s4\}$. For this scenario, the total achieved network lifetime is 3 units. Therefore, we can easily observe that avoiding a fraction of targets (one target in this case) in each cover set can considerably improve functional duration.

Researchers in the past have addressed many heuristics to provide solutions for partial coverage under WSNs [16, 24, 25]. The authors in [24] first represented the partial target coverage (known as α-coverage) in the form of a linear program and further proved it to be a *NP*-complete problem because partial coverage is a special case of full coverage. Since the research work in [12] proved the full coverage problem to be a *NP*-complete problem, partial coverage is also a *NP*-complete problem. The same study [24] further presented a thorough methodology for finding cover sets of partial coverage problems, where sensors with the highest remaining energy and those covering critical targets are preferred first while generating cover sets. Raiconi et al. [25] are the first ones who addressed partial coverage using genetic algorithms, where cover sets are represented as chromosomes and are formed through where crossover techniques. While generating these cover sets, this study also considered the coverage and residual energy of the critical target in deployed networks. The authors in [16] improved network lifetime compared to the studies in [24, 25] by generating minimal cover sets while keeping alive those sensors that were covering critical target(s). Along with residual energy and coverage of the critical target, this is prime concern. By keeping sensors that cover critical targets alive for a longer period, the authors tried to achieve the optimal upper bound of the network lifetime. Further, the same heuristic avoids critical targets in the cover set to fulfil the partial coverage requirement. This way, one can certainly improve the network's functional duration.

3.4.4 Energy-Efficient Connected Coverage

Merely collecting data from the deployed sensor network is useless unless the same data can reach the central processing location, or BS, for future usage to achieve different outcomes based on the need of the application. This considers both the coverage and connectivity of the sensors with the base station [17, 18]. Since the deployed network is dense, many sensors within that network are far from the base station and are not connected directly to it. In such cases, data is transferred from one sensor to the adjacent 1-hop neighbour sensors until it reaches the BS. The authors in [17] proposed an algorithm based on 1-hop neighbour packet forwarding, where nodes send data to their nearest neighbour and continue doing so until the packet is not received by the BS. In order to select a node with multiple 1-hop neighbours, nodes with highest residual energy are chosen first to extend network lifetime. Furthermore, a similar approach is followed by the research work in [18], where residual energy as well as the degree of each node is considered to forward the data to the nearest neighbour so that packets can reach the BS. We have discussed distinct variants of coverage. The main aim behind all these variants is to extend network functional duration by applying various heuristic approaches. Now, in the next section, we will discuss many heuristics and their comparative analysis on the achieved network lifetime, which are based on the metaheuristic approach. Before a comparative analysis, we will discuss the background of genetic algorithm-based metaheuristics for target coverage variants.

3.5 GENETIC EVOLUTIONARY METAHEURISTIC

Nowadays, genetic algorithm-based metaheuristics are widely applied for providing solutions to optimization problems [26–29]. Here, we first discuss the basic operation and functionality of genetic algorithms so that readers can get acquainted with the working methodology of the metaheuristics, which are devised with the concept of genetic algorithm characteristics. Therefore, we start this section with a detailed background on it.

3.5.1 INTRODUCTION TO GENETIC ALGORITHM-INSPIRED EVOLUTIONARY TECHNIQUES

Almost all metaheuristic algorithms are inspired by biological evolutionary techniques. Genetic algorithms are the major basis of these evolutionary methodologies, where various growth concepts of genetic algorithms are applied while designing these metaheuristics. A lot of work exists on genetic algorithms in the literature [6, 26–30]. The behaviour of the chromosomes in the population is reflected in the performance of genetic algorithms. Here, chromosomes are binary strings where a particular digit/bit is called a gene. The population is nothing but the collection of these chromosomes over a period. Since we are studying evolutionary techniques with respect to target coverage, here chromosomes represent cover sets. Further, the size of the chromosome is decided by the number of sensors in the deployed networks. When considering cover sets represented as chromosomes, the digit 1 indicates that the respective sensor is part of the cover set, and 0 indicates that the corresponding sensor is not part of the cover set. Initially, randomly generated cover sets are considered as the population, which is known as the search space of genetic algorithms. Each and every chromosome is assigned a fitness score, which simply shows how competent the chromosomes are in the population.

All the evolutionary techniques to optimize energy consumption in networks are composed of three basic steps, namely, selection of chromosomes, crossover to generate new chromosomes from the parent chromosomes, and mutation to convert invalid chromosomes into valid chromosomes. During the selection phase, two parent chromosomes are randomly picked from the initial population pool, which are further responsible for producing child chromosomes. Once this phase is over, the next task is to perform a crossover to produce the next generation of child chromosomes. Lastly, to improve the fitness of the chromosomes generated in the crossover phase, a mutation operation is performed to enhance the proficiency of the chromosomes. Now we will discuss the basic operation followed by these evolutionary techniques to solve the target coverage in WSNs, where network dynamics are constantly changing over time. Previous studies have made significant progress with the help of evolutionary methodologies.

TABLE 3.9

Sensors and Target Coverage

s/t	t_1	t_2	t_3	t_4
s_1	1	0	0	0
s_2	1	0	1	0
s_3	0	0	1	1
s_4	1	1	0	1
s_5	1	0	0	1
s_6	1	0	0	1
s_7	1	0	1	1

	S_1	S_2	S_3	S_4	S_5	S_6	S_7
Sensors →							
Gene →	1	0	1	1	0	0	0

FIGURE 3.7　Chromosomes representing cover set

3.5.1.1　Generating Chromosome

As we discussed earlier, chromosomes, CG, represent cover sets, which are combinations of $0s$ and $1s$ and can be defined as below in (3.8).

$$CG_i = \begin{cases} 1, if \text{ sensor } s_i \text{ is active} \\ 0\, otherwise \end{cases} \tag{3.8}$$

Illustration 3

To understand the structure of chromosome CG, let us consider the following network given in the form of a matrix a shown in Table 3.9, where there are seven sensors and four targets spread over the same terrain, and every sensor is assigned 1 unit battery of life. The network is considered to be homogeneous. Here, we can observe the network and find that $s1$, $s3$, and $s4$ form a cover with four targets with 1 unit of battery life.

Now, in order to represent sensor cover using a chromosome, we define the chromosome, CG, as follows. Figure 3.7 shows that the chromosome has the same number of sensors (i.e. seven) as given in the network shown in Table 3.9. Further, one of the cover sets of the network given in Table 3.9 has three sensors, namely $s1$, $s3$, and $s4$, and hence the chromosome of the respective cover set has three 1s and the rest of the digits are all 0s, as shown in Figure 3.7.

After knowing the structure of the chromosome for the deployed network, our next task is to form an early population pool of chromosomes where a fixed number of chromosomes are kept for basic operations such as crossover.

```
Input: X (Number of sensors), S (pool size)

Output: S (Number of chromosomes)

    • bool CG [S]
    • init Pool ={}
    •     for i=1 to S
    •         for j= 1 to X
    •             CG [j] = random()%2
    •         end for
    •        Pool = Pool ? CG
    •     end for
```

FIGURE 3.8   Early population (pool) generation

3.5.1.2 Population Generation

At first, the early population pool has chromosomes that are simply a string of 0s and 1s which may or may not be a cover set for the deployed networks. To generate these random strings of 0s and 1s, we follow the procedure as shown in Figure 3.8. Here, one inbuilt function random () is used which helps us generate sequences of 0s and 1s in a random manner. Once the early population pool is generated, one can proceed to the next step of genetic algorithms, where chromosomes are first picked from this pool only. It is clearly shown in Figure 3.8 that chromosomes for the early population pool are generated by randomly created 0s and 1s.

As we already discussed, the size of the chromosome is decided by the network size, which refers to how many sensors are in the deployed network. Further, it can be observed that the chromosome generated here is a random sequence of 0s and 1s, and it may not necessarily be a cover set. In the next selection phase, steps are taken to make sure that the selected chromosome represents a valid cover set.

3.5.1.3 Selection Operation

Once the early population pool is formed in the previous stage, the next operation is to select the valid chromosomes (i.e. those representing a cover set) from the pool. Here, a valid chromosome represents a cover set. Thus, the selection phase ensure that only those chromosomes that represent cover sets are chosen. In order to perform this selection operation, various methods have been followed by researchers, including tournament, rank, and roulette-wheel selection [31]. Based on the given network scenario or requirement, one can choose any of these methods for selection of valid chromosomes.

3.5.1.4 Crossover

Genetic algorithm-inspired heuristics are adapted from biological evolution concepts, and new chromosomes are created by selecting two parent chromosomes and then performing a crossover operation in order to improve the quality of the parent

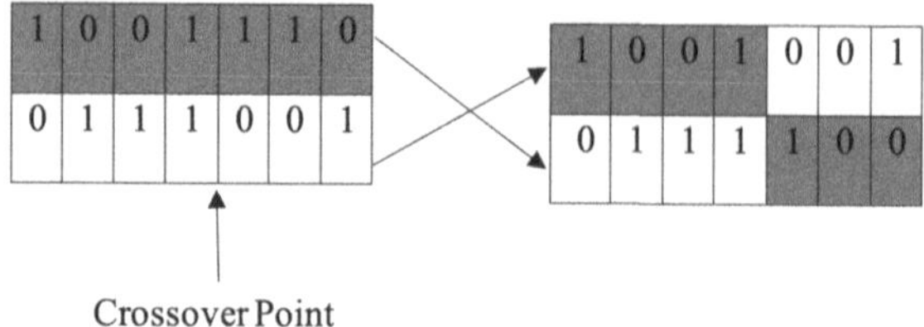

FIGURE 3.9 One-point crossover

chromosomes. With respect to the coverage scenario, quality means how important the cover set is while enhancing the total functional duration of deployed WSNs. Therefore, the child chromosomes (i.e. newly generated chromosomes) are better than the parent chromosomes in terms of prolonging the lifetime of the networks. In the literature, there are many crossover techniques, including one-point, two-point, and uniform crossover [32]. To understand the functionality of one-point crossover, Figure 3.9 can be referred to. This shows the functionality of one-point crossover where gene values are exchanged after the crossover point for the parent chromosomes to generate a child chromosome.

Figure 3.9 shows that two parent chromosomes, one in grey and another in while (left side of Figure 3.9), are chosen form the pool at the selection phase. After performing a one-point crossover operation on the left side of the two parents, we achieve two child chromosomes (on the right side of Figure 3.9) by swapping bits after the crossover point. Thus, the crossover operation is performed successfully. At the end of this operation, we can observe that two more new chromosomes are formed with the help of the two parent chromosomes which were initially picked up from the early population pool.

3.5.1.5 Mutation

During the crossover operation, child chromosomes are created. By doing so, the resultant child chromosome may be an invalid chromosome. As discussed earlier in this section , a valid chromosome represents a sensor cover in the case of the target coverage problem. Therefore, a mutation operation is needed to convert the invalid chromosome into a valid one by flipping the bits (i.e. gene value) from 0 to 1 or vice versa [33].

Illustration 4

To understand necessity of the mutation operation, consider Table 3.10 for the genetic algorithm-inspired coverage metaheuristics in a wireless sensor network. Here, we have considered five sensors and four targets, and a homogeneous network where each sensing device is equipped with a unit battery with the same sensing range.

Let us choose two parent chromosomes (both are valid) for this network from the pool of early population to generate two child chromosomes using a one-point crossover operation, as depicted in Figure 3.10, to create two child chromosomes after crossover operation is over.

TABLE 3.10

Sensor Network

s/t	t_1	t_2	t_3	t_4
s_1	0	1	1	1
s_2	1	0	1	0
s_3	1	0	1	1
s_4	0	0	1	0
s_5	1	1	0	1

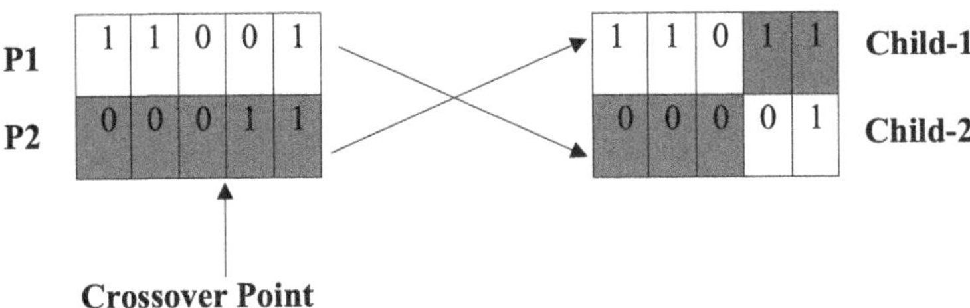

FIGURE 3.10 Mutation operation

Child-1 represents a valid chromosome as it is a sensor cover, but this is not the case for Child-2 as it does not represent a valid chromosome because this is not a cover set. In order to make Child-2 a cover set (i.e. a valid chromosome), we need to perform a mutation operation which will flip a few 0s to 1s and make it a cover set.

After presenting a detailed discussion on the functionality of genetic algorithms, one can easily understand the working of metaheuristics for solving target coverage in WSNs. In the next section, we will discuss a few existing metaheuristics for solving target coverage.

3.5.2 ENERGY-EFFICIENT COVERAGE WITH GENETIC METAHEURISTICS

Target coverage heuristics are used to find cover sets over time until network is functional. After every iteration, sensor batteries are depleted for those in the current cover set, which in turn changes the deployed network dynamics over time, and genetic algorithm-based methodologies are best suited for such applications. Since genetic algorithms are evolutionary techniques and mostly applicable to optimization problems, target coverage can be solved efficiently. In this section, we will discuss a few existing algorithms for solving variants of the coverage problem and also discuss their comparative analysis on the achieved network lifetime [18, 27–29].

Before an in-depth discussion of these existing heuristics mentioned by authors in [18, 27–29], we should understand the concept of the fitness function and its importance while designing metaheuristics for target coverage problems. The chromosome's quality is typically decided by the fitness function.

3.5.2.1 Metaheuristics Functionality

The main phases of all the metaheuristics can be represented as shown in Figure 3.11, which are generally followed. Here, in Figure 3.11, we observe that one important step of any metaheuristic is the fitness function calculation. Most of the metaheuristics differ in the way they calculate fitness functions. Next, we will discuss some important metaheuristics based out of genetic algorithms to prolong the functional duration of the WSNs with respect to target coverage [18, 27–29]. All the existing methodologies first differ in the way they design their fitness functions and, second, how they perform the rest of the operations followed by the genetic algorithms.

Next, we will present a comprehensive comparative analysis of a few of the existing metaheuristics inspired by genetic algorithms.

3.5.2.2 Metaheuristic Comparative Analysis

The authors in [18] discussed a genetic algorithm-based heuristic that has a fitness function with three criteria for providing energy-efficient coverage and connectivity. The discussed metaheuristics provide solutions for target K-coverage as well as M-connectivity where sensors are connected with a M number of neighbouring sensors all the time. Thus, the addressed heuristic is suitable for target full coverage (if we set $K = 1$), target K-coverage, and connected coverage too. While choosing sensors for the cover set, the authors made sure that the fewest sensors cover all the

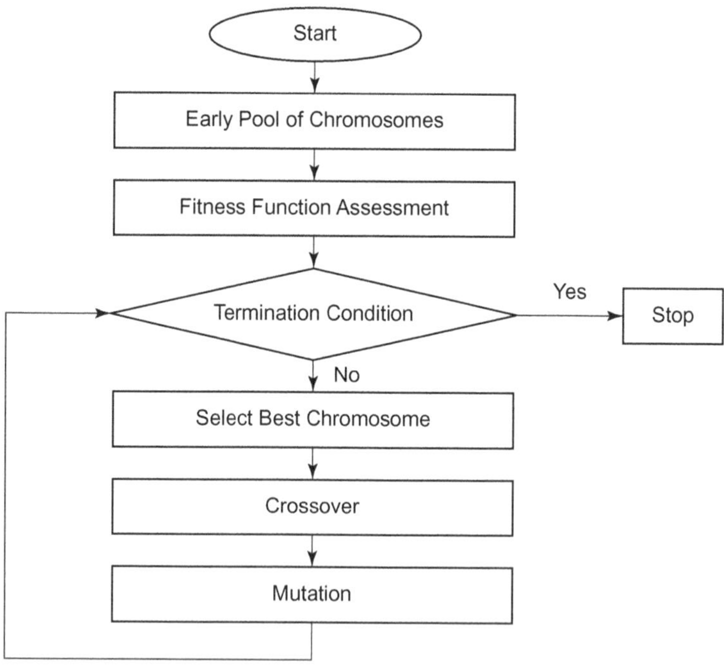

FIGURE 3.11 Flow of metaheuristic

targets. If we need to avoid connectivity, we can consider one aspect of the discussed heuristics by setting $M = 1$.

The authors in [27] experimented with 20 different genetic algorithm-based methods on three energy models in order to identify which genetic algorithm-based method is most suitable for which energy model. The discussed work presented detailed simulation results with a vast set of input combinations of genetic algorithms and energy models to extend network functional duration. In order to minimize energy consumption while communicating with the base station, one has to select the recommended combination of genetic algorithms and energy models.

Harizan et al. [28] discussed another metaheuristic for coverage and connectivity in WSNs. The discussed methodology basically focuses on mutation operation customization for prolonging lifetime. To modify the mutation operation, the authors made sure that any sensor not part of coverage and connectivity was removed from the cover set. By doing so, sensors can be alive for a longer time and in turn can be part of future sets, thus allowing the network to last for an extended period. The discussed study was also compared with existing works, where the authors proved that their proposed methodology outperformed previous methods of coverage and connectivity for coverage problem in WSNs.

Recently, the authors in [29] addressed an altogether new metaheuristic for target coverage in WSNs. The addressed proposal has a two-fold fitness function and modified mutation operation which directly impacted network life. Here, the fitness function ensured the selection of the fewest sensors that cover critical targets. The objective behind such selection is to ensure that these types of sensors have remaining batteries for an extended period, which in turn delays energy hole development in the networks. As already discussed, critical targets become uncovered first in the deployed networks; hence, keeping alive critical sensors will certainly improve the performance of the proposed methodology for target coverage. Furthermore, the modified mutation operation also adds an extended functional period to these critical sensors by wisely flipping bits from 0 to 1. Generally, the mutation operation randomly flips bits from 0 to 1 to make the chromosomes valid. However, this metaheuristic does not perform the usual mutation operation; instead, a modified mutation is applied, which further extends the network's functional duration in WSNs. Additionally, the metaheuristic addressed in [29] outperformed the studies discussed and proposed in [27, 28].

3.5.2.3 Simulation Setup and Experimentations

So far, we have discussed many metaheuristics based on genetic algorithms that improve the total achievable network lifetime in WSNs [27–29]. In this section, we want to explore these metaheuristics in more detail and will also perform a comparative analysis. To perform the experiments, we need to decide on many network parameters and specifications.

For the simulation, we considered a heterogeneous network spread over a squared region with two types of dimensions, namely 200*200 m^2 and 300*300 m^2. Sensors have a sensing range equal to 100 m, and we also assume a randomly generated network where sensor nodes, targets, and X and Y coordinates are generated using a pseudorandom generation technique. The simulation was carried out on MATLAB® (2016), and an initial pool of generations consists of 60 chromosomes which are randomly generated with the help of the algorithm given in Figure 3.8. The mutation rate is 3%. After 150 fixed iterations, all the metaheuristics stop calculating network lifetime for the given network configurations.

Experiment 1

As we have already discussed, metaheuristics mainly differ in the way they calculate their respective fitness functions. The metaheuristics discussed in the previous section generally have two- or three-fold fitness functions. To understand this, let us consider the metaheuristic discussed in [29] where majorly two objectives are there to form cover sets.

Here, we consider only one metaheuristic [29] for an explanation and better understanding of the concepts of evolutionary techniques.

Objective 1: (Select least critical sensor coverage)

So far, we have mentioned many times that the impact of critical target coverage massively affects achievable network lifetime because these are the targets whose coverage stops first, and thus the network becomes non-functional once their coverage ceases. The sensors covering such critical targets are of prime concern, and hence, the first objective of metaheuristics [29] is to minimize these sensors in a cover set. Let us say that $X_{Critical}$ are these sensors and X is the total sensors, which gives the following expression:

$$Minimize\ F1 = \frac{X_{Critical}}{X} \tag{3.9}$$

Objective 2: (Minimize the generated cover set)

Once critical targets are covered with the help of $X_{Critical}$ sensors, the remaining sensors should participate in the minimum required number. None of the targets should have redundant coverage. Therefore, this objective tries to minimize the generated cover set to extend the network's functional duration). Let us say that $X_{minimal}$ sensors are needed more for covering leftover targets, then the second objective can be expressed by the following expression:

$$Minimize\ F2 = \frac{X_{minimal}}{X_{Critical}} - X \tag{3.10}$$

The fitness function designed in metaheuristics [29] is constructed by summing up both the objectives in a weighted sum approach [34]. Here, all the objective functions are separately multiplied by a weight (W). For the aforementioned two objective functions, the fitness function can be presented as follows:

$$\textbf{Fitness} = \text{W1} \times (1-\text{F1}) + \text{W2} \times (1-\text{F2}) \tag{3.11}$$

$$\textbf{Objective}: \text{Maximize fitness} \tag{3.12}$$

This method is generally known as the weighted sum approach for genetic algorithms, where fitness function comprises more than one parameter. The weighted sum approach proved that $W1 + W2 = 1$, where $0 \leq Wi \leq 1$, $\forall i, 1 \leq i \leq 2$. All these metaheuristics based on genetic algorithms claim that the chromosome quality directly depends on the fitness function. Therefore, there are many possible combinations of $W1$ and $W2$ which may lead to the best fitness function. The authors in [29] experimented with a few combinations of $W1$ and $W2$, and their results are shown in Table 3.11 and Table 3.12 for different network dimensions with the same network configuration for the rest of the parameters.

In this simulation, various values of $W1$ and $W2$ are considered, which are between 0 and 1. To conduct this experiment, we have considered 50 targets, and sensors vary between 50 and 250. Further, the deployed network is considered to be heterogeneous, and we experimented with two different network dimensions, namely, 200*200 m^2 and 300*300 m^2. The authors in [34] mentioned that if the problem area is familiar, then finding combinations of weights is quite possible. To do so, we experimented with 30 combinations of $W1$ and $W2$, and out of those, five possible combinations outperform the rest.

As depicted in Table 3.11 and Table 3.12, the metaheuristic addressed in [29] with the weights $W_1 = 0.6$ and $W_2 = 0.4$ achieves a reasonable lifetime when compared with the other four combinations of weights. Hence, throughout the entire simulation setup, experiments are carried out with $W1 = 0.6$ and $W2 = 0.4$ weight values. Here, these five combinations are considered to be $C1 = \{W1 = 0.6, W2 = 0.4\}$, $C2 = \{W1 = 0.5, W2 = 0.5\}, C3 = \{W1 = 0.4, W2 = 0.6\}, C4 = \{W1 = 0.45, W2 = 0.55\}$, and $C5 = \{W1 = 0.35, W2 = 0.65\}$.

It is clearly depicted in Table 3.11 and Table 3.12 that $W_1 = 0.6$ and $W_2 = 0.4$ weight values can achieve better network lifetime compared with the rest of the weight combinations. In other experiments, we consider these two values of weight to enhance the total network lifetime for the deployed networks.

Experiment 2

Here, we discuss the performance of various metaheuristics designed and proposed by various authors [27–29] for prolonging the functional duration of the deployed networks. To do so, we have considered 50 targets, and sensors vary between 50 and 250 in each iteration. Here, we experimented with two network scenarios, namely 200*200 m^2 and 300*300 m^2, and the obtained results are shown

---

TABLE 3.11

Network Lifetime under Area 200*200 m^2

Sensors	C1	C2	C3	C4	C5
50	8	8	7	7	6
100	15	13	14	12	10
150	31	28	28	25	23
200	36	33	31	29	27
250	45	42	41	40	36

TABLE 3.12

Network Lifetime under Area 300*300 m²

Sensors	C1	C2	C3	C4	C5
50	8	7	6	5	5
100	12	11	11	10	8
150	28	26	25	22	21
200	33	30	28	24	25
250	42	40	38	35	33

TABLE 3.13

Network Lifetime under Area 200*200 m²

Sensors	Harizan [28]	Sunil [27]	Manju [29]
50	6	6	8
100	12	11	15
150	28	26	31
200	33	32	36
250	41	40	45

in Table 3.13 and Table 3.14. During deployment, sensors are assigned battery units between 1 to 2 units randomly.

The network lifetime achieved by various existing metaheuristics [27–29] is shown in Table 3.13 where sensors are vary with a fixed set of targets (50). Here, we observe two major findings from the data given in the above table. First, the network functional duration extends with more sensors. This happens because network size is fixed, and targets to be covered are also fixed. With more sensors under the same fixed underlying network, the same set of targets will be covered for a longer period which in turn increases the network lifetime achieved. The second finding is that the performance of the proposed metaheuristic addressed by authors [29] is better than those heuristics mentioned in [27, 28]. This happens due to the fitness function and the mutation operation chosen by these heuristics in different ways. As already discussed, all metaheuristics based on genetic algorithms differ in the way they choose their fitness functions and mutation operations. This is reflected in Table 3.13. The same experiment is repeated with different network dimensions (300*300 m²), with the rest of the network parameters remaining the same. Experimental results are depicted in Table 3.14.

Here, again, Table 3.14 depicts two major findings. First, the network functional duration exponentially increases according to the number of sensors, and the metaheuristic proposed in [29] outperforms the other studies mentioned in [27, 28]. Apart from these two findings, we can also observe that the network lifetime achieved in

TABLE 3.14

Network Lifetime achieved under Area 300*300 m^2

Sensors	Harizan [28]	Sunil [27]	Manju [29]
50	4	4	5
100	11	10	12
150	26	24	28
200	31	30	33
250	40	39	42

Table 3.14 is less than that mentioned in Table 3.13 for a particular set of sensors and targets. This is because in Table 3.13, network size is 200*200 m^2 and the data in Table 3.14 is for a network size of 300*300 m^2. Due to the same set of sensors and targets but a larger deployed region, more sensors are required to provide coverage of the targets, which are randomly deployed initially. Due to this, fewer sensor covers are generated, which in turn decreases the network lifetime for the deployed region. The same findings can be demonstrated in Figure 3.12 and Figure 3.13 with this network configuration, where the deployed network is assumed to be heterogeneous, and sensors are randomly assigned battery units between 1 and 2 units.

As depicted in Figure 3.12, lifetime is prolonged with more sensors, and total achieved network lifetime according to the study in [29] is more than that achieved by the metaheuristics in [27, 28]. This is because these metaheuristics have different fitness functions and mutation operations. Now, the same experimental result is shown in Figure 3.13 for a different network size from Figure 3.12 where the network size is taken as 300*300 m^2 instead of 200*200 m^2. Here, again, we can observe the same outcomes regarding network lifetime dependency on the sensors and the size of the deployed network. Further, the total network lifetime achieved by the metaheuristic in [29] is better than that achieved by the studies in [27, 28] with the same network configurations.

Experiment 3

Here, we demonstrate the performance of the metaheuristics designed and proposed by various authors [27–29] for prolonging the total network lifetime for the deployed networks. To do so, we have considered 250 sensors, and targets vary between 50 and 200 in each iteration. Here, we simulated both networks, namely 200*200 m^2 and 300*300 m^2, and the obtained results are shown in Table 3.15 and Table 3.16.During deployment, sensors are assigned battery units between 1 to 2 units randomly.

The network lifetime achieved by various existing metaheuristics [27–29] is shown in Table 3.15, where sensors are fixed at 250 and targets vary between 50 and 200. Here, we observe two major findings from the data. First, the functional duration of the network decreases with more targets. This happens because network size is fixed and the sensors covering these increasing targets are also fixed. With increasing targets in the same fixed area, the same set of sensors covering these

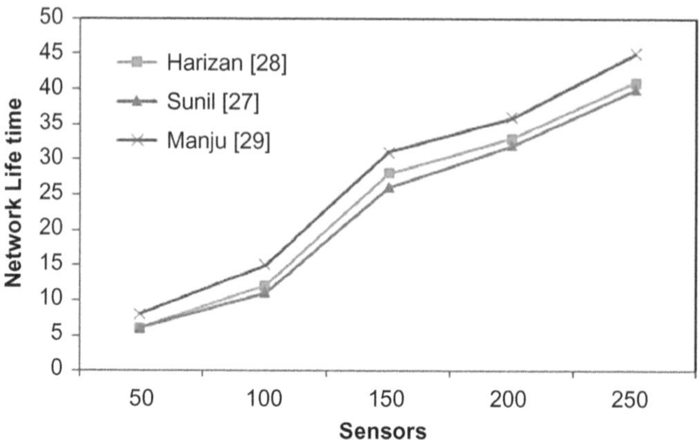

FIGURE 3.12   Network lifetime achieved in the sensing area 200*200 m^2

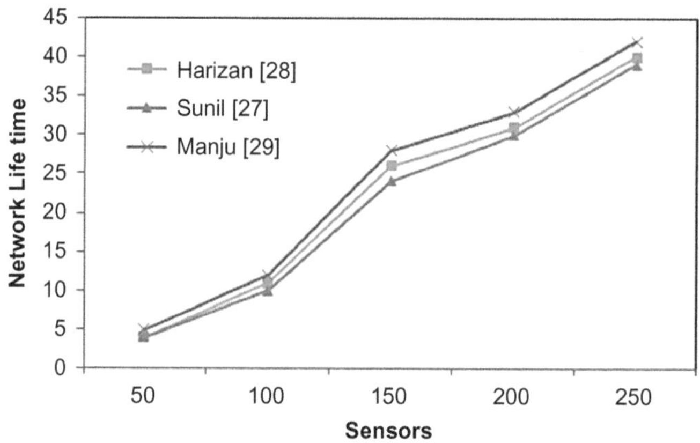

FIGURE 3.13   Network lifetime achieved in the sensing area 300*300 m^2

targets results in fewer cover sets, which decreases the total network functional duration. The second finding is that the performance of the metaheuristic mentioned in [29] is better than those heuristics mentioned in [27, 28]. This happens due to the fitness function and the mutation operation chosen by these heuristics in different ways. As we already discussed, all the metaheuristics based on genetic algorithms differ in the way they choose their fitness functions and mutation operations. This is reflected in Table 3.15.

The same experiment is repeated with different network dimensions (300*300 m^2), with the rest of network parameters being the same. The experimental outcomes are shown in Table 3.16. Here, again, Table 3.16 depicts two major findings. First, the network functional duration decreases with more targets, and the metaheuristic proposed in [29] outperforms the rest of the studies mentioned in [27, 28]. In addition,

TABLE 3.15
Network Lifetime under Area 200*200 m^2

Targets	Harizan [28]	Sunil [27]	Manju [29]
20	52	51	54
40	47	45	48
60	38	37	40
80	28	26	32
100	20	19	23

TABLE 3.16
Network Lifetime under Area 300*300 m^2

Targets	Harizan [28]	Sunil [27]	Manju [29]
20	48	47	51
40	44	43	46
60	32	30	35
80	22	21	25
100	11	10	13

we can observe that the network lifetime achieved in Table 3.16 is less than those mentioned in Table 3.15 for a particular set of sensors and targets. This is because in Table 3.15, network size is 200*200 m^2 and Table 3.16 data is for a network size of 300*300 m^2. Due to the same set of sensors and targets but a larger deployed region, more sensors are required to cover the same targets randomly deployed. Due to this, fewer sensor covers result in a decreased lifetime for the deployed region.

The same findings are shown by Figure 3.14 and Figure 3.15 with the above-discussed network configuration, where the deployed network is assumed to be heterogeneous and sensors are randomly assigned battery units between 1 and 2 units.

As depicted in Figure 3.14, the network lifetime decreases with more targets, and the total achieved network lifetime according to the study in [29] is more than that achieved by the metaheuristics in [27, 28]. This is because these metaheuristics have different fitness functions and mutation operations. The same experimental result is shown in Figure 3.15 for a different network size from Figure 3.14, where the size is 300*300 m^2 instead of 200*200 m^2.

Here, again, we can observe the same outcomes regarding network lifetime dependency on the targets and the size of the deployed network. Further, the total functional duration attained with metaheuristics [29] is better than the studies in [27, 28] with the same network configurations. So far, we have discussed many variants of the target coverage problem in WSNs. We have experimented with target full coverage in detail and compared many metaheuristics that are inspired by the concept

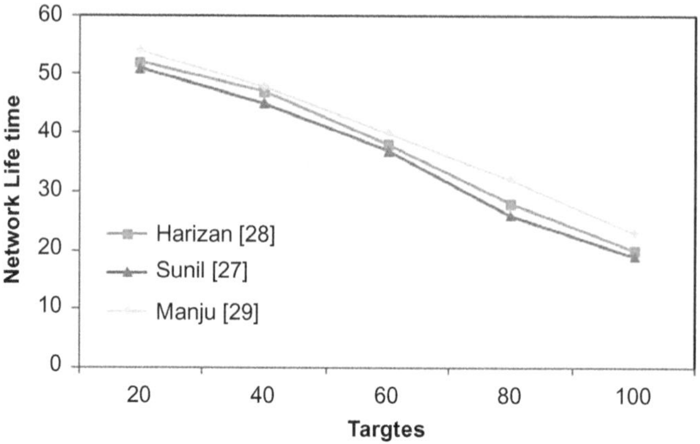

FIGURE 3.14   Network lifetime achieved in the sensing area 200*200 m²

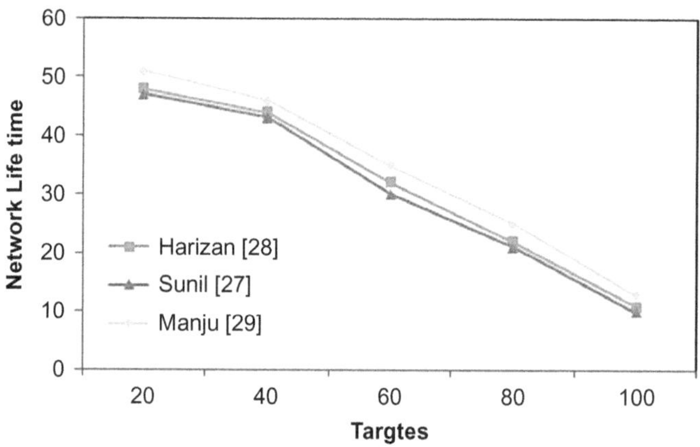

FIGURE 3.15   Network lifetime achieved in the sensing area 300*300 m²

of the genetic algorithm's paradigm. We also simulated for the various sensing areas, changing sensors and targets in the same deployed network to see how network functional duration is affected by these parameters.

3.6 CONCLUSION

In WSNs, there are two major coverage problems known as target coverage and area coverage. This chapter addresses the target coverage problem where targets are deployed in a pre-defined terrain along with the sensor nodes. The objective of these sensing devices is to monitor these targets for the longest possible duration, known as network lifetime. Many variants of target coverage are addressed in the literature, namely target full coverage, target K-coverage, target partial coverage, and target connected coverage. Targets are considered covered as long as they are covered by at least one sensor. Hence, the scarce resources of the network (battery) is the prime concern, impacting the functional duration of the network. The sensors have a limited energy source (battery life) which is non-rechargeable as well as non-replaceable depending upon the network application. Therefore, one has to optimize the usage of the sensor battery in the correct way to prolong the network lifetime for any type of deployed network. Further, genetic algorithm-based metaheuristics are proven to be best suited for solving the target coverage problem variants, and hence we have discussed a few of these heuristics [27–29] and given a detailed explanation of how genetic algorithm-based metaheuristics work for solving coverage concerns in WSNs. We have performed detailed simulations to show their comparative performance analysis on the deployed networks of different dimensions. We have experimented with varying sensors and targets with different energy levels and observed that the heuristics addressed in the study in [29] outperforms the rest of the existing works [27, 28].

REFERENCES

1. Akyildiz, F., W. Su, Y. Sankarasubramaniam, and E. Cayirci. "A Survey on Sensor Networks." IEEE Communications Magazine 40, no. 8 (2002): 102–14.
2. Dargie, W., and C. Poellabauer. *Fundamentals of Wireless Sensor Networks: Theory and Practice*. New York: John Wiley and Sons, 2010.
3. Gowrishankar, S., T. G. Basavaraju, D. H. Manjaiah, and S. K. Sarkar. "Issues in Wireless Sensor Networks," In Proceedings of the World Congress on Engineering, London, UK, vol. 11, pp. 176–87, 2008.
4. Liang, Jubin, Ming Liu, and Xiaoyan Kui. "A Survey on Coverage Problems in Sensor Networks." *Sensors & Transducers* 163, no. 1 (2014), 240–6.
5. Kotzanikolaou, P., C. Douligeris, D. Zorbas, and D. Glynos. "BGOP: An Adaptive Algorithm for Coverage Problems in Wireless Sensor Networks." 13th European Wireless Conference, EW2007, 2007.
6. Zorbas, D., D. Glynos, P. Kotzanikolaou, and C. Douligeris. "Solving Coverage Problems in Wireless Sensor Networks Using Cover Sets." *Ad Hoc Networks* 8 (2010), 400–15.
7. Chong, C.Y., and Kumar, S. P. "Sensor Networks: Evolution, Opportunities and Challenges," *In Proceedinfs of the IEEE* 91, no. 8 (2003): 1247–56.
8. Harb, Hassan, Abdallah Makhoul, Rami Tawil, and Ali Jaber. "Energy-Efficient Data Aggregation and Transfer in Periodic Sensor Networks." *IET Wireless Sensor System* 4 (2014): 149–58.
9. Jubin Liang, Ming Liu, and Xiaoyan Kui. "A Survey on Coverage Problems in Sensor Networks." *Sensors & Transducers* 163, no. 1 (2014), 240–46.
10. Nguyen, T. G., C. So-In, and N. G. Nguyen. "A Novel Energy-Efficient Clustering Protocol with Area Coverage Awareness for Wireless Sensor Networks." *Peer-to-Peer Network Application* 10 (2017), 519–36.
11. Manju, Anuradha. "A Novel Energy-Efficient Heuristic for Target Coverage to Maximize Sensor Network Lifetime." *International Journal of Computer Applications (IJCA)* 86, no. 7 (January 2014): 31–5.
12. Cardei, M., M. T. Thai, Y. Li, and W. Wu. "Energy-Efficient Target Coverage in Wireless Sensor Networks." In Proceedinds of IEEE Infocom, 2005.

13. Manju, S. Chand, and B. Kumar. "Maximizing Network Lifetime for Target Coverage Problem in Wireless Sensor Networks." *IET Wireless Sensor System* 6, no. 6 (December 2016): 192–7.

14. Jiguo Yu, Shengli Wan, Xiuzhen Cheng, Dongxiao Y. "Coverage Contribution Area Based k-Coverage for Wireless Sensor Networks." IEEE Transactions on Vehicular Technology, 2017, vol. 66

15. Manju, Samayveer Singh, Sandeep Kumar, Anand Nayyar, Fadi Al-Turjman, Leonardo Mostarda. "Proficient QoS-Based Target Coverage Problem in Wireless Sensor Networks." *IEEE Access* 8 (2020): 74315–25.

16. Manju, Chand, S. and Kumar B. "Selective α-Coverage Based Heuristic in Wireless Sensor Networks." *Wireless Personal Communications* 97 (2017): 1623–36.

17. Yang, C., and K. W. Chin." On Nodes Placement in Energy Harvesting Wireless Sensor Networks for Coverage and Connectivity." *IEEE Transactions on Industrial Informatics* 13, no. 1 (2017): 27–36.

18. Gupta, S. K., P. Kuila, and P. K. Jana. "Genetic Algorithm Approach for K Coverage and M-Connected Node Placement in Target Based Wireless Sensor Networks." *Computers and Electrical Engineering* 56 (November 2016): 544–56.

19. Mini, S., S. K. Udgata, and S. L. Sabat. "Sensor Deployment and Scheduling for Target Coverage Problem in Wireless Sensor Networks." *IEEE Sensors Journal* 14, no. 3, (2014): 636–44.

20. Pujari, A. K., S. Mini, and T. Padhi. "Polyhedral Approach for Lifetime Maximization of Target Coverage Problem." *ICDCN* (2015): 14:1–14:8. https://doi.org/10.1145 /2684464.2684495.

21. Chaudhary, M. and A. K. Pujari, "Q-Coverage Problem in Wireless Sensor Networks," In *Proceedings Int. Conf. Distrib. Comput. Netw. ICDCNin (Lecture Notes in Computer Science)*, vol. 5408. Berlin, Germany: Springer-Verlag, 2009, pp. 325–330

22. Manju, Pawan Bhambu, and Sandeep Kumar. "Target K-Coverage Problem in Wireless Sensor Networks." *Journal of Discrete Mathematical Sciences and Cryptography* 23, no. 2 (2020). https://doi.org/10.1080/09720529.2020.1729511.

23. Yu, J., Y. Chen, L. Ma, B. Huang, and X. Cheng, "On Connected Target k-Coverage in Heterogeneous Wireless Sensor Networks," *Sensors* 16, no. 1 (2016): 104.

24. Gentili, M., and A. Raiconi. "Alpha-Coverage to Extend Network Lifetime on Wireless Sensor Networks." *Optimization Letters* 7, no. (2013): 157–72.

25. Cerulli, Francesco, Carrabs Raffaele, Ciriaco D'Ambrosio, and Andrea Raiconi. "A Hybrid Exact Approach for Maximizing Lifetime in Sensor Networks with Complete and Partial Coverage Constraints." *Journal of Network and Computer Applications* 58 (2015): 12–22.

26. Konak, A., D. W. Coit, and A. E. Smith. "Multi-Objective Optimization Using Genetic Algorithms: A Tutorial." *Reliability Engineering & System Safety* 91 (2006): 992–1007.

27. Sunil Kr. Jha, and Egbe Michael Eyong. "An Energy Optimization in Wireless Sensor Networks by Using Genetic Algorithm." *Telecommunication System* 67 (2018): 113–21.

28. Harizan, S., and P. Kuila. "Coverage and Connectivity Aware Energy Efficient Scheduling in Target Based Wireless Sensor Networks: An Improved Genetic Algorithm-Based Approach." *Wireless Networks* 25 (2019): 1995–2011.

29. Manju. "A Meta-Heuristic Based Approach with Modified Mutation Operation for Heterogeneous Networks." *Wireless Personal Communication* 122 (2022): 963–79. https://doi.org/10.1007/s11277-021-08935-w.

30. Katoch, S., S. S. Chauhan, and V. Kumar. "A Review on Genetic Algorithm: Past, Present, and Future." *Multimedia Tools Applications* 80 (2021): 8091–126.

31. Zhou, Jian, and Zhongsheng Hua. "A Correlation Guided Genetic Algorithm and Its Application to Feature Selection." *Applied Soft Computing* 123 (2022). https://doi.org/10.1016/j.asoc.2022.108964.
32. Aspiro, S. and S. Karagol. "An Experimental Study of the Effect of Different Genetic Algorithm Operators on Coverage Using Heterogeneous Nodes." 2022 International Symposium on Multidisciplinary Studies and Innovative Technologies (ISMSIT), Ankara, Turkey, 2022, pp. 619–23.
33. allah Mottaki, Nemat, Homayun Motameni, and Hosein Mohamadi. "A Genetic Algorithm-Based Approach for Solving the Target Q-Coverage Problem in Over and Under Provisioned Directional Sensor Networks." *Physical Communication* 54 (2022): 2022. https://doi.org/10.1016/j.phycom.2022.101719.
34. Konak, A., D. W. Coit, and A. E. Smith. "Multi-Objective Optimization Using Genetic Algorithms: A Tutorial." *Reliability Engineering & System Safety* 91, no. 9 (2006): 992–1007.

Chapter 4

SUMMARY

In this chapter, we address connectivity and communication issues in the field of the wireless sensor networks. First, we address connectivity in WSNs and then the associated issues concerned with energy-optimized connectivity where the major challenge is to efficiently utilize the limited battery associated with the sensor nodes in the deployed networks. Once connectivity is ensured, then communication protocols are selected wisely depending on the application size and proximity. Further, there are many challenges that impact the selection of these communication protocols to ensure energy conservation. We address the connected coverage in detail where the objective is to monitor a certain set of targets and then make sure that the sensors that are monitoring these targets are connected to the base station as well forwarding the collected information. In order to provide connected coverage, we discuss many reinforcement learning-based metaheuristics, which are mainly based on learning automata. We also compare their performance with detailed simulation results. The experiments check their performance on connected coverage with varying sensors, targets, and sensing ranges.

DOI: 10.1201/9781003427780-6

4 Connectivity and Communication in Wireless Sensor Networks

4.1 INTRODUCTION: WIRELESS SENSOR NETWORKS

Wireless sensor networks (WSNs) are comprised of numerous small sensing devices called sensor nodes. A sensor node generally communicates with neighbouring nodes to share or receive data to/from the central master node, which is commonly known as the base station (BS). For communication and operation, sensors are embedded with a small inbuilt internal memory where users can store relevant information, a microprocessor which makes the sensor able to operate collected data, and a small onboard battery to make the sensor operational until the battery exhausts [1]. In this chapter, we will discuss the connectivity and communication of these sensing devices to pass the gathered data to remaining sensors or the central BS in order to further process the collected data in the deployed WSNs. In the next sections, we present a detailed discussion on communication protocols, various network paradigms, connectivity in WSNs, and various issues related to connectivity and communication [2].

4.2 CONNECTIVITY AND COMMUNICATION IN WSNS

Once the WSNs are deployed, the main focus of the deployed network is to gather information from all the sensing devices. There are many types of sensors to collect specific types of data in the deployed networks. For example, there are temperature sensors to measure the temperature of the surroundings where it is deployed, humidity sensors to record average humidity in the given sensors' sensing range terrain, motion sensors to record the mobility of the objects in the surroundings, pressure sensors to measure the pressure of the surroundings, vibration sensors, and many more [3, 4]. Thus, there is a specific sensor for a particular task, and to get that data, the respective sensor has to send that data to either the neighbouring sensors or the BS where further decisions are made. To communicate with adjacent nodes, there are various communication protocols that are widely used in various applications

DOI: 10.1201/9781003427780-8

depending on suitability and requirements [5]. In the next section, many connectivity- and communication-related issues are discussed in deployed wireless sensor networks.

4.3 CONNECTIVITY ISSUES IN WIRELESS SENSOR NETWORKS

Once the network deployed for the respective activity, the major issue behind deployed WSNs is how well these sensing devices are connected to each other for communicating gathered data or information to each other. Thus, most of the methodologies differ in the way they choose the next hop node to connect with the rest of the network. There are many works in the literature that are concerned with how to connect these nodes so that energy consumption is less in order to make the rest of the network connected [6–9]. Once the network becomes connected in a certain way depending on the applied connectivity protocol, one has to ensure that these connected nodes communicate with each other in order to exchange information and forward it to the base station [10]. Before going into the details of these communication protocols, we present various connectivity paradigms for the network to be connected. Numerous factors affect the connectivity of the given WSNs, including single-hop or multi-hop architecture, topology of the network, and the type of the network. Now we will discuss all these issues one by one in the following sections.

4.3.1 TYPE OF ARCHITECTURE

The connectivity of the deployed nodes is still a major problem, as sensors can be connected to each other via a single hop (directly connected to the other nodes) as shown in Figure 4.1 or multiple hops (with the help of intermediate nodes) as shown in Figure 4.2. This is possible because generally the deployed networks are dense in nature, where sensor nodes are deployed in very close proximity to each other. It can be observed in Figure 4.1 that each node of the deployed network is directly connected to the base station to communicate any type of data or information collected over time from its surroundings; hence, this is known as single-hop architecture.

In contrast to this connectivity, the network in Figure 4.2 depicts nodes that are far from the BS and are not connected to it directly; instead, they are connected via intermediate nodes (hop nodes), which is hence referred to as multi-hop architecture. Further, a hop node can be an intermediate node for multiple nodes deployed in the same network if connecting node is not placed in close proximity. Hence, the architecture plays an integral role in connecting the nodes of the network either with neighbouring nodes or to the BS for information exchange purposes.

The major challenge while connecting via single-hop architecture is the network infrastructure itself. There is a chance that not all nodes are directly connected to the base station in the deployed networks. In this situation, we need to consider multi-hop architecture. However, there are other challenges with this scenario, including intermediate node failure or delays in sending data to the BS due to multiple intermediate nodes. Thus, connectivity type should be chosen wisely based on the

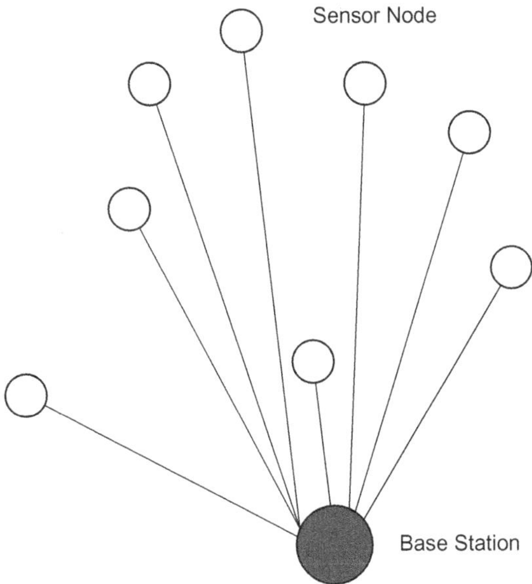

FIGURE 4.1   Single-hop architecture

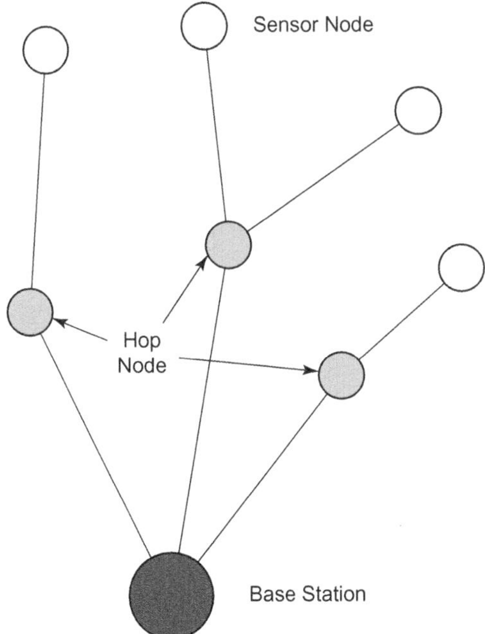

FIGURE 4.2   Multi-hop architecture

application requirements. For an energy-optimized connectivity paradigm, one has to make an intelligent choice here; otherwise, the network become dysfunctional at an early stage due to the limited energy associated with every single node of the network.

4.3.2 Type of Topology

Once the network is deployed, the participating nodes get connected to each other in different topologies: star topology, bus topology, mesh topology, ring topology, and tree topology. Now we present an overview of these connectivity topologies to understand the issues related to connectivity in networks [11]. ***Star*** topology refers to the type of connectivity where sensor nodes are directly connected to a central node and should not connect to each other. As shown in Figure 4.3, none of the network nodes is connected to the rest of the nodes except the central node. Here, nodes form a shape like a star; that is why it is generally known as star topology.

The advantage of the star topology is that the complete data gathered can be found in one place, the central node. The main disadvantage is the single-path dependency issue, where if a node's connection with the central node fails, there is no other way to connect that node to the remaining network to transfer the collected data. Again, this type of connectivity is not always possible, so one has to study the application requirements properly in order to deploy the network. This way, the star topology can be opted for while ensuring connectivity of sensing devices with the BS. ***Ring*** topology refers to a type of the connectivity where sensor nodes are connected in such a way that each node is connected to two neighbouring nodes at a time, as shown in Figure 4.4.

The major advantage of ring topology over the star topology is that nodes are connected to two neighbouring nodes, and hence there is no issue of single node failure as compared to the star topology. However, the disadvantage here is that the same data or information is communicated to two nodes by each node of the network

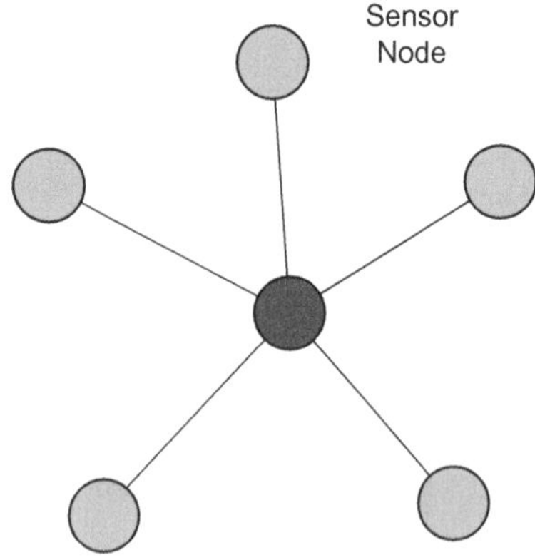

FIGURE 4.3   Star topology

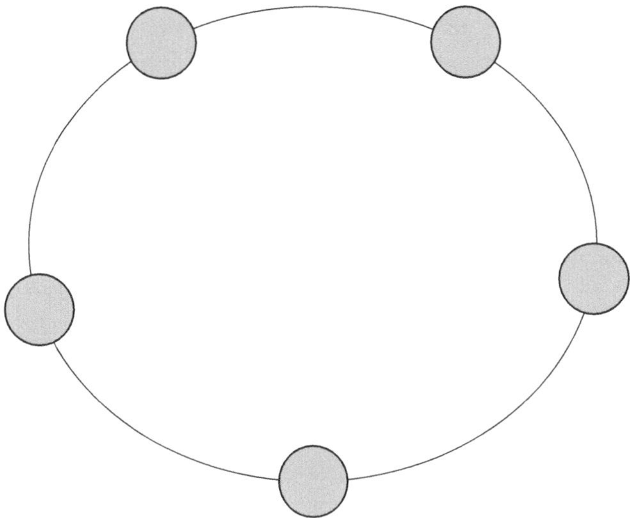

FIGURE 4.4 Ring topology

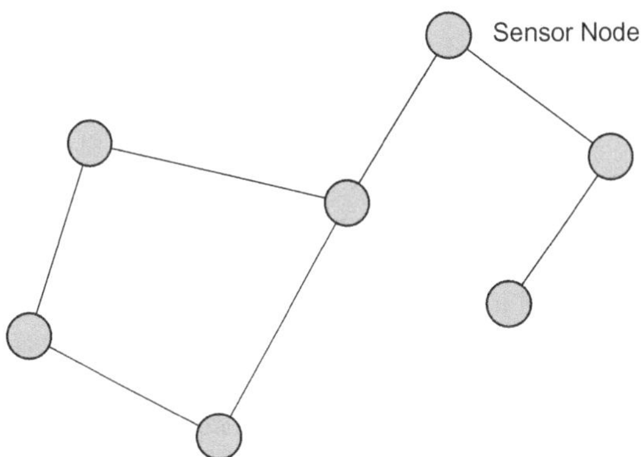

FIGURE 4.5 Partial mesh topology

in every iteration. As already discussed, the network should be energy optimized. Therefore, while choosing a type of connectivity paradigm, one has to be aware of energy usage to ensure the network does not fail.

Mesh topology refers to the type of connectivity where a sensor is connected to many other neighbouring nodes so that they can forward the collected data from the surroundings to the base station. In general, partial mesh and full mesh are prominent technologies. Partial mesh topology has connectivity of nodes as shown in Figure 4.5, and a full mesh topology ensures that each node is directly connected to the rest of the network, as shown in Figure 4.6. Hence, we can observe that the nodes can choose any path to send the data to the BS. The advantage of such topologies is

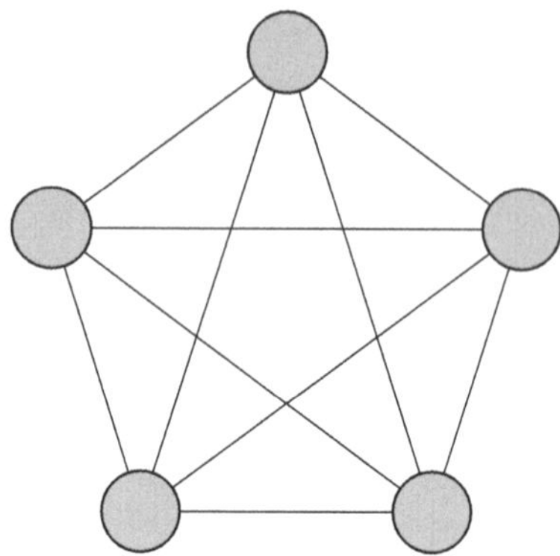

FIGURE 4.6 Full mesh topology

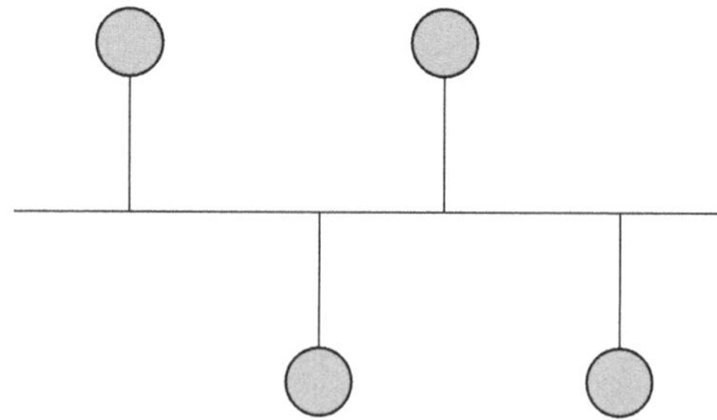

FIGURE 4.7 Bus topology

that sensor nodes are able to share the collected data to the BS, and there is no fear of path failure.

When comparing both mesh topologies, one can face altogether different challenges based on the application requirements, which should be studied well to choose the optimal topology for better network connectivity.

Depending on the type of deployed network, the respective topology is adopted. If the deployed network is dense and nodes are quite close to each other, one can adopt a fully connected network architecture; otherwise, one can choose partially meshed architecture. In both cases, one can face altogether different challenges while providing connectivity in the deployed network. ***Bus*** topology refers to the connectivity of nodes where nodes are connected over a common path, and they are not connected to each other directly, as shown in Figure 4.7. If a particular node has to send data to some other node, then the sender node simply broadcasts the message over the bus channel, and only the intended receiver node receives the data. While using bus topology, one can face the challenge associated with single-path failure, as all the nodes are connected via a single central bus path. Therefore, one must choose wisely chose when considering the connectivity of nodes in the deployed networks.

These are some commonly used topologies for the deployed network, which are adopted by the participating sensor nodes to connect with each other in the same networks. Next, we discuss the various types of networks that also affect node connectivity in transferring information to the base station.

4.3.3 Type of Network

As we know, WSNs are extremely application-specific. Therefore, the type of network deployed for a particular application may have different connectivity issues. In order to explain various connectivity issues involved with specific types of networks, we need to explain these types first and then their associated connectivity issues. According to existing research, there are mainly four types of networks: static and mobile WSNs, homogeneous and heterogeneous WSNs, deterministic and non-deterministic WSNs, and single-base station and multi-base station WSNs. Here we present a detailed discussion of all of these network types and explain their respective connectivity issues [12–14].

4.3.3.1 Static and Mobile WSNs

The deployed network may be static or dynamic based on the application requirements. In the case of static networks, sensor nodes cannot change their location from where they are initially placed throughout the functional duration, whereas in mobile networks, sensors can relocate between various locations within the given network proximity. When a network is static, any type of topology can be used to place these nodes, but in a mobile network, these topologies change throughout the process based on the current network connectivity requirements. Therefore, mobile networks have complex connectivity issues compared to static network because sensors are moving between locations, and thus, connectivity concerns increase. On the other hand, static networks are quite stable, and once the network is deployed, there is no movement, which in turn helps to provide better connectivity for the underlying networks.

4.3.3.2 Homogeneous and Heterogeneous WSNs

While deploying the sensor network, it is important to know whether the application requires a homogeneous or heterogeneous network. The choice of selecting the type of network solely depends on the requirements of the deployed network. In the case of homogeneous networks, sensors are assigned equal sensing range as well as battery life. All network configurations are the same in this type of coverage. In contrast, heterogeneous networks have sensors that are equipped with different sensing ranges and batteries to cope with application requirements. So, the choice of selecting either of these types of sensor networks directly impacts connectivity issues. Certainly, homogeneous and heterogeneous networks have different types of connectivity issues. In order to know the challenges in advance when providing connectivity, one has to ensure that the correct type of network is chosen.

4.3.3.3 Deterministic and Non-Deterministic WSNs

A deterministic WSN is the one where we know the geographic locations of the sensors to be deployed in advance, as shown in Figure 4.8. There are applications of WSNs where the network is humanly accessible and requires a limited number of sensors. In such cases, deterministic deployment takes place, as we know the locations of the sensors to be deployed. However, most deployed WSNs are not humanly accessible, for example in a wildlife monitoring application where humans cannot access the terrain in person. In such scenarios, non-deterministic deployment is required where sensor nodes are just spread over the region by aircraft, as shown in Figure 4.9. Here, we can observe that nodes are positioned in random locations in the designated area.

When we compare Figure 4.8 and Figure 4.9, we can observe that deterministic deployment has a fixed structure and nodes are placed in these fixed locations, whereas in non-deterministic deployment, sensors are randomly spread over the given region. Thus, depending on the application area requirements, one has to choose between them, which results in different connectivity issues.

There are fewer connectivity issues in deterministic deployment due to advanced location information, but connectivity issues inevitably arise in non-deterministic deployment. To connect with adjacent nodes, these sensors must first locate them in the network by following some distance calculation methods and then identifying nodes that fall within their sensing ranges. Since it takes time to establish connectivity, many connectivity issues arise in the deployed networks. Thus, the choice of network type will certainly impact the performance of the deployed network while providing connectivity in the underlying networks.

4.3.3.4 Single-Base Station and Multi-Base Station WSNs

During WSN deployment, it is required to keep a base station within the network where sensors can send their collected data, as sensors are equipped with less

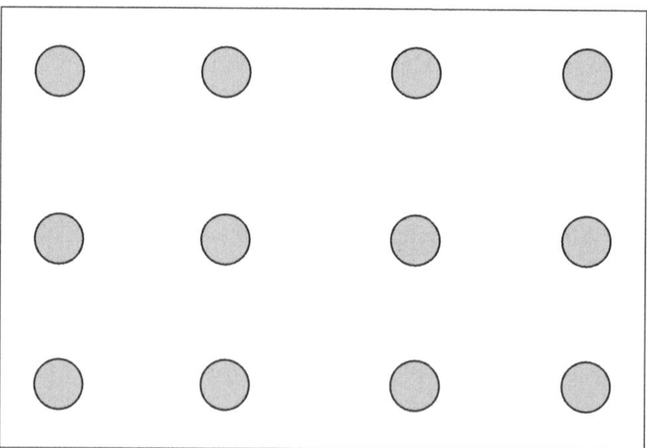

FIGURE 4.8 Deterministic deployment

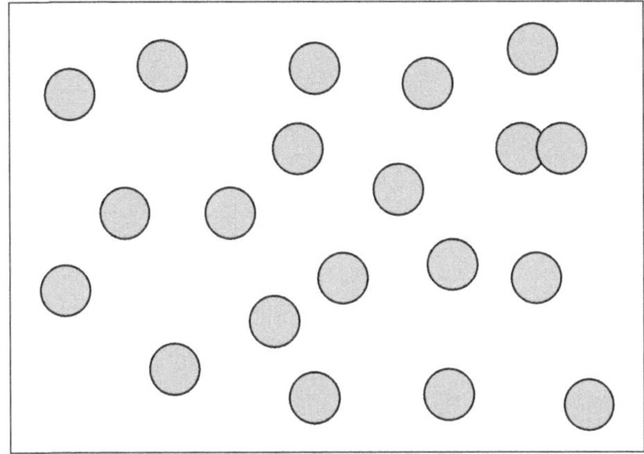

FIGURE 4.9   Non-deterministic deployment

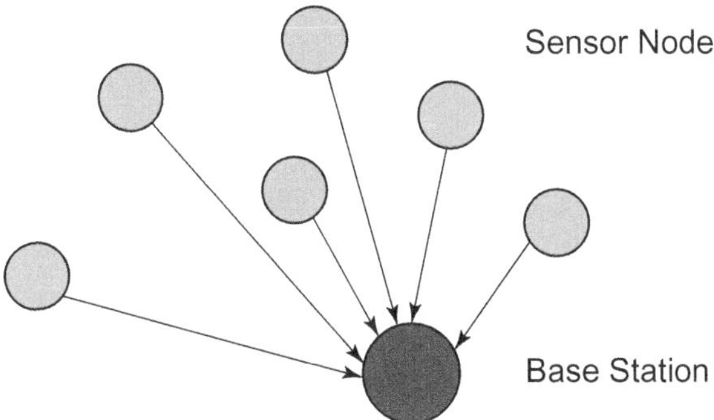

FIGURE 4.10   Single-base station deployment

memory and hence cannot store much data. So, irrespective of the type of network deployed, there should be at least one base station within the network, as shown in Figure 4.10. Here, all the deployed sensors can send their data to the given base station.

There are some applications where the sensor network is deployed over a large geographic region, and a single-base station will create too much delay while sensors are sending data to it. In such cases, there must be one or more base stations in the same underlying network so that collected data can be transferred speedily, as shown in Figure 4.11. Here, there are two base stations, which increase network performance while providing connectivity with reduced latency.

As shown in Figure 4.11, there are two base stations in the same network, and sensors send data to the nearest BS to speed up the process. Therefore, depending on the type of network, connectivity issues arise. In the case of single-base stations, all the sensors should transfer data to the only BS, while in the case of multiple base

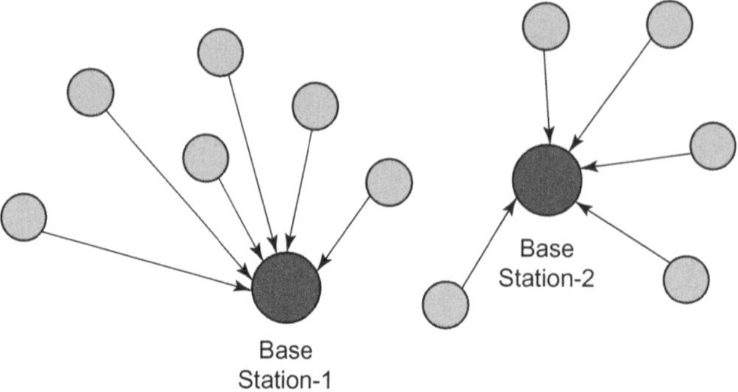

FIGURE 4.11   Multi-base station deployment

stations, there must be some rules and guidance for the deployed sensors so that they can be connected to one of the base stations only.

So far, we have addressed many connectivity issues concerning various network paradigms, which includes static and mobile WSNs, homogeneous and heterogeneous WSNs, deterministic and non-deterministic WSNs, and single-base station and multi-base station WSNs. In the next section, we will discuss various communication paradigms for wireless sensor networks and their issues while communicating data between nodes in the same or different networks. These major communication paradigms are basically divided based on their communication ranges and the type of technology they use for communication with the other nodes of the network.

4.4   COMMUNICATION PARADIGMS IN WIRELESS SENSOR NETWORKS

Once the deployed network ensures that all the sensors are connected to the concerned neighbouring nodes under a certain topology, communication between these nodes becomes the next challenge. To communicate, the connected sensors follow a particular communication paradigm based on the requirements fixed by the underlying network architecture and the application. In the literature, there are many communication protocols that can be adapted by the deployed network to enable communication between the participating nodes [15–18].

Before a detailed discussion of these communication protocols, we need to understand some critical factors affecting the choice of communication protocol for the chosen application, which will be given in the sections that follow [19].

4.4.1   Communication Issues in WSNs

There are certain issues concerned with the choice of communication protocol in the deployed wireless sensor network. The following paragraphs provide details of a few of the issues.

Limited energy: Sensors are equipped with small batteries that are non-rechargeable. In addition, sensors often cover terrain that is sometimes not humanly accessible, meaning that the power sources are non-replaceable. Because of this, we need to be careful while selecting communication paradigms to conserve these limited energy sources and prolong network functional duration. If the selected protocol is not energy-efficient, it will make the underlying network dysfunctional at a very early stage.

Hardware constraints: As we already discussed, sensors are tiny devices that come with small hardware units, which primarily include an internal memory and processor. With small processors and memory, sensors are not able to store and process huge data for long periods. Therefore, this limitation should be taken into account when selecting a communication paradigm so that the collected data can be forwarded as soon as possible without the loss of a single bit of information. By doing so, one can ensure guaranteed delivery of collected data to the BS.

Node deployment: In general, the node deployment strategy can certainly impact the communication efficiency of the network. If nodes are randomly deployed, then it become crucial to maintain proper communication throughout the network's functional duration. Deterministic deployment somewhat minimizes this problem, as node location is known in advance, which can make it easier for communication protocols to function.

Dynamic network: Frequently changing the network topology can certainly be a challenge when providing communication between the deployed nodes. This becomes crucial when the network is mobile and sensors can relocate within the network. In such situations, the selected communication protocol should be capable enough to cope with the changing dynamics of the deployed networks.

Fault tolerance: Network communication should be operational under any infrastructural situation, which includes sensor node failure, base station failure, or connectivity failure. These communication protocols should be designed to quickly adapt to the changing environment.

Latency: This is the measure of time taken to forward the data to the concerned destination node. Whichever communication protocols are chosen,they should not increase the latency beyond a certain threshold. The communication latency is also known as end-to-end delay when calculated for the deployed wireless sensor networks.

Scalability: There are many applications where network deployment starts with a few sensor nodes and over time keeps growing to meet application requirements. In such scenarios, the communication protocols should be well equipped to meet the growing demands. Further, the network performance should not degrade with the growing size of the underlying network.

4.4.2 COMMUNICATION PROTOCOLS

Now, we will discuss the existing communication protocols designed for WSNs in detail. To exchange data between sensors, sensing devices should communicate with each other. There are many communication protocols addressed in the literature based on the type of application chosen [20]. These communication technologies are mainly classified as short-distance wireless communication technology, medium-distance wireless communication technology, and long-distance communication technology [21, 22]. We will discuss each of these categories in detail in the following section of the chapter so that we can choose the most suitable one based on requirements.

4.4.2.1 Short-Distance Communication Technology

In WSNs, radio-frequency identification (RFID) along with Bluetooth are generally used as communication protocols [23]. These short-distance communication protocols can communicate within a range of 10 m at most. This type of communication is required when the deployed network is mostly indoors.

Bluetooth is a communication protocol used to exchange data or information between devices (be it fixed or mobile) that are close to each other at very short distances. Bluetooth uses ultra high frequency (UHF) waves, which are mainly radio waves, and operates from 2.402 GHz to 2.480 GHz. While communicating, Bluetooth forms personal area networks (PANs) to forward data from one device to other devices in the PANs [24]. Bluetooth uses a radio technology called frequency-hopping spread spectrum. To exchange data, Bluetooth generally divides the information packet into several packets of the same or different sizes and then transmits these prepared packets on one of the 79 designated Bluetooth channels. Here, all the channels have 1 MHz bandwidth. In general, there are 1600 hops per second, with adaptive frequency-hopping (AFH) enabled. Bluetooth uses 2 MHz spacing, which accommodates around 40 channels. Bluetooth is a low-power consumption device and hence mostly used in WSNs as a communication protocol [25].

Radio-frequency identification (RFID) technology generally uses radio waves, which operate at several different frequencies while transferring data. RFID is mainly comprised of two types of components, namely *tags* and *readers*. In general, a *reader* is a device which not only keeps emitting radio waves but also receives signals coming back from the RFID tags. A *tag* also uses radio waves to communicate its identity and other information to the nearby *readers* in the same proximity. RFID tags are known as passive or active tags. Active RFID tags are equipped with batteries to become operational, while passive RFID tags are powered by the reader and do not have any built-in power source to make them functional [26]. Thus, Bluetooth and RFID technologies are mainly used for communication in the deployed WSNs for short-range communication. If the deployed network is wider than this, one can chose from other options, which is discussed in the following section.

4.4.2.2 Medium-Distance Communication Technology

In order to maintain communication between two sensing devices, Wi-Fi [27] and Zigbee [28] are used for medium-distance information exchange. These medium-distance communication protocols can communicate within a range of 10 m–100 m at most. To perform this type of communication, we have two methods, as discussed below.

Wi-Fi (Wireless Fidelity) is among the most popular wireless sensor network communication protocols, which works for the wireless local area network (WLAN). Generally, Wi-Fi protocol follows the IEEE 802.11 standard, operating through 2.4 GHz frequency. In WSNs, sensor devices that are 20–40 meters away from the source can access the Internet via Wi-Fi. Based on the choice of antenna and the channel frequency used, the Wi-Fi protocol has a data rate of up to 600 Mbps maximum [27]. Generally, Wi-Fi is not commonly used in WSNs due to higher power consumption and less data range, as sensor nodes are spread over a large geographic region [29].

Another medium-distance communication protocol for WSNs is **Zigbee**, which has almost the same features as Bluetooth technology. Unlike Bluetooth, Zigbee follows the IEEE 802.15.4 standard and can communicate over a range of 10–100 meters at 250 Kbps. It also has low-power consumption similar to Bluetooth, robustness, and extended high security, as well as scalability, which makes it a suitable choice for communication in WSNs. Due to the limitation on range, it is suitable for small-scale wireless sensor networks applications. Apart from that, it also provides 1258–bit AES (Advanced Encryption Standard) to secure transmission of data over devices [30]. Thus, depending on the underlying architecture, users can adopt any of the above-mentioned protocols.

4.4.2.3 Long-Distance Communication Technology

To communicate over long distances, well-known cellular networks (2G/3G/4G) and low power wide area (LPWA) are mainly used. These long-distance communication protocols can communicate to devices over 100 m away [28]. Over the last two decades, the cellular network has been chosen as a long-distance communication protocol in WSNs. It is mainly comprised of GSM/GPRS/EDGE(2G)/UMTS or HSPA(3G)/LTE(4G) communication protocols. With the help of this cellular network, a large amount of data can be transferred over long distances. The cellular network protocol operates at frequencies from 900 to 2100 MHz, and it covers distances of 35 km to 200 km. The operational speed of the cellular network is generally high, from 35 Kbps to 10 Mbps.

Low power wide area (LPWA), sometimes referred to as mobile IoT (Internet of Things), is a network communication paradigm that transmits data at slower rates than other communication protocols, especially cellular technologies. Low power wide area also costs less, consumes less power, and provides improved area coverage in rural and underwater areas, or inside buildings, unlike other cellular technologies. The LPWA protocol was not widely used until IoT, especially with mobility, became prevalent. With the help of LPWA, these connected moving devices are better managed when it comes to transmitting and receiving the data [31].

So far, we have addressed numerous types of communication protocols that are available for WSN devices to communicate with neighbouring nodes. To do so and exchange information, these sensing devices must have connectivity between them. We have addressed many communication protocols for forwarding data between nodes or the base station. In the next section, we will address the well-known target connected coverage in WSNs, where coverage and connectivity with the base station is ensured.

4.5 CONNECTED COVERAGE AND REINFORCEMENT METHODOLOGY

In order to provide connectivity in the deployed wireless sensor networks, the participating sensor nodes should be connected to each other as well as with the base station [32]. In this chapter, we are studying connectivity with respect to target coverage; as discussed in Section 4.2 of this chapter. Next, we will discuss the connected coverage and then provide a detailed introduction of reinforcement methodologies, which are mainly inspired by learning automata, for solving connected coverage in the field of WSNs.

4.5.1 Introduction to Connected Coverage

Merely providing coverage is not enough until the data or information by nodes participating in the current sensor cover is transferred to the BS for further processing. To do so, nodes in the cover set must be connected to each other as well as to the base station. To achieve this, intermediate nodes, sometimes called hop nodes, play an important role. To provide connectivity between nodes of the current cover set and the BS, hop nodes are chosen wisely so that the total energy consumed by these active nodes remains minimal. Energy-efficient connected coverage ensures both target coverage and connectivity while consuming less energy, as the network is energy constrained [33]. The connected coverage is a special variant of target coverage where, after finding a cover set for the deployed network, one must ensure that all the participating sensors in the cover set are connected to the BS so that they can collectively forward gathered data to BS [34]. Being a special case of target coverage, connected coverage is also a NP-complete problem. There are various ways to solve this problem, and next, we will address the learning-automata-based reinforcement methodology for connected coverage in WSNs. The major objective of applying the reinforcement learning-based paradigm is due to the dynamic behaviour of the network over time. Before going into details, we will address how the reinforcement methodology-inspired learning-automata works for optimization problems such as connected coverage in the underlying WSNs.

4.5.2 Overview of Reinforcement Methodologies

WSNs operate in a dynamic environment where the characteristics of the network change with time due to various factors, which include mobility of the nodes, mobility of the base station, sensor node transition between active and sleep modes, and node battery drainage. For these reasons, the network dynamics change and impact the performance of any energy-efficient connected coverage protocols. This unavoidable fact can certainly impact the choice of protocol that can be adopted in the changing network dynamics [35].

In recent years, reinforcement learning (RL) has received increasing interest for coverage and connectivity issues aimed at improving the network performance based on artificial intelligence [36]. Reinforcement learning is an adaptive online learning approach. This online approach makes it possible for any protocol to acquire knowledge spontaneously whenever environmental dynamics change.

The reinforcement learning approach is based on unsupervised and online learning techniques. This type does not need an external observer to monitor the learning process; instead, there is a decision-maker, known as the agent, which attempts to understand the surroundings. With the help of online learning, the agent acquires knowledge as required, and hence empirical data is not required. In general, there are mainly three representations that are embedded in each agent of reinforcement learning. These representations are given below [37].

1. The state of an agent represents the prime decision-making factors considered by the agent from the dynamically changing surroundings. These states affect the action selection for the next iteration and the rewards (i.e. network performance). In the field of WSNs, a state can be the remaining battery of a sensing device and the size of a buffer queue, which is equal to the count of packets in it.
2. *Action* is a notion of an agent's action which is responsible for changing or affecting the reward (i.e. network performance) or state (i.e. operating environment). Therefore, the agent here spontaneously learns from the environment to take the optimal action in most cases. In the case of connected coverage, the actions are considered as the selection of transmission power or the selection of the next hop node, which will further transmit the packet towards the base station or next hop nodes.
3. *Reward* basically represents the action results or impacts of the previous action on the state in terms of the performance of the underlying networks. The rewards can be a gain or loss achieved by the agent after taking certain action on its current state. During connected coverage in the WSNs, the rewards are energy consumption level by a node or throughput achieved by all the participating nodes. With the help of trial-and-error, the long-term rewards by each pair of state-action are determined by the reinforcement learning in the operating environment. The workflow of the reinforcement learning is shown in Figure 4.12.

The learning engine component is the prime and most vital element in Figure 4.12. During this whole process, state and reward are observed by the agent from its operating environment. Based on these observations (learning), the agent decides the appropriate action to be carried out in the environment in such a way that, for the next iteration, state and reward can be improved. Here, there may be the possibility that the action taken by the agent may affect the state and reward for better or for worse, or maintain the status quo; this in turn affects the agent's selection the next time.

The application of reinforcement learning in WSNs can be considered where the learning phase is to find the optimal path to the BS from the sensing device of the cover set. Here, a state can be the destination node, or base station, and an action is considered the choice of the next hop node to make connectivity. Penalty and reward

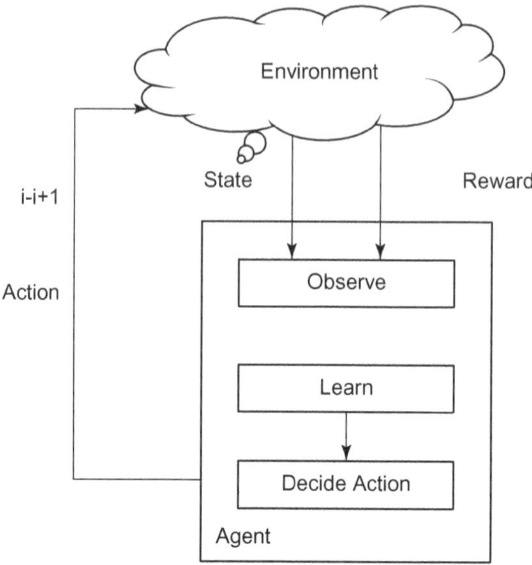

FIGURE 4.12   Simplified reinforcement learning model

can be defined as the length of the path from the cover set nodes to the BS. If the selected path has a shorter distance, then it is treated as a reward; if it is longer than the previous one, then it will be treated as a penalty. If the action is rewarded, it will certainly improve network performance.

4.5.3   Connected Coverage Using Reinforcement Learning Automata

In the previous section, we discussed how reinforcement learning works for optimization problems where the objective is to maximize connectivity by continuously learning from the environment. Now, we will discuss well-known connected coverage in WSNs, where we aim to cover the deployed targets for the maximum duration and ensure that the sensors connected to the base station participate in sensor cover [38, 39]. To achieve this, we address learning-automata-based metaheuristics for solving connected target coverage. We have already mentioned that the changing environment of the wireless sensor network is conducive to adapt such reinforcement learning methodologies. Before a detailed discussion, we need to address the functionality of learning automata in detail to understand why it is suitable for the connected target coverage problem in deployed WSNs.

4.5.3.1   Learning Automaton (LA)

Learning automata (LA) is known as reinforcement learning based on artificial intelligence and is frequently used in WSNs where there is a need to adapt to the characteristics of the changing environment, which in turn would result in a significant increase in network performance [40]. Before a detailed discussion on connected coverage using learning automata, this section first addresses related details.

In general, "learning" refers to a process of gaining knowledge to modify one's behaviour to respond to specific actions in the given environment. There is a collection of actions in an automaton, and for each of these actions, the environment either responds in favour or may go against the action with some probability. Therefore, one can say that the automaton plays an important role as decision-maker for actions that are based on a fixed set of rules to attain the best output. In this overall process, an LA keeps interacting with the surroundings in a continuous manner so that it can make the best decisions by selecting the appropriate action in each iteration of the chosen metaheuristic.

In general, the learning-automata environment, LE, is the medium where learning-automata functions are mainly represented as a tuple $LE=\{a,c,b\}$, where a, b, and c are defined as follows [41]:

$a=\{\alpha1,\alpha2,...,\alpha r\}$ are finite inputs,

$b=\{\beta1,\beta2,...,\beta r\}$ are finite outputs, and

$c=\{c1,c2,...,cr\}$ are penalty probabilities where $ci\varepsilon c$ is for each αi

Initially, every automaton can choose an action αi from the given action set $\{\alpha1,...,\alpha r\}$ along with a penalty probability ci . As we discussed earlier, LA's are pre-assigned with an action probability vector that corresponds to their initial set of actions. Further, the action probability essentially drives the learning automata by reward/penalty depending on final output achieved in the last iteration. By observing these reward-penalty assignments, one can help the LA determine what future action to take in the given environment.

Figure 4.12 clearly shows the learning functionality of a learning automaton. In general, learning automata are mostly known as fixed and variable structure LA. In fixed structure LA, the action probability vector is static, while in variable structure LA, it changes over time. Since the network environment is constantly changing over time in wireless sensor networks, the variable structure LA is best suited for this study. The action selection by any automaton from the action pool completely depends on the updated action probability vector, which keeps evolving based on the applied reward-penalty operation. Since variable structure LA depends on the learning algorithm, which is responsible for the changes in the action probability vector, variable structure LA can be represented as a quadruple $\{\alpha,\beta,P,H\}$. Here we know that $\alpha=\{\alpha1,...,\alpha n\}$ is a set of initial actions, and $\beta=\{\beta1,...,\beta n\}$ is defined as a set of all the inputs. Further, $P=\{P1,...,Pn\}$ is defined as a set of action probabilities and H as the learning algorithm, which is later used by the automaton. To understand the complete process performed by an automaton, we need to consider the following steps to be performed one by one in sequence.

Step 1: With the given set of actions, $\alpha=\{\alpha1,...,\alpha n\}$, the automaton randomly selects an action αi based on the action probability p to perform in the given environment LE.

Step2: After receiving the reinforcement signal, the automaton updates its action probability vector. In order to update these action probabilities for favourable and unfavourable actions, refer to (4.1) and (4.2) as defined in [41].

$$pi(n+1)=pi(n)+a\left[1-pi(n)\right]$$

$$pj(n+1)=(1-a)pj(n)\,\forall j, j\neq i \qquad (4.1)$$

$$pi(n+1)=(1-b)pi(n)$$

$$pi(n+1)=b/r-1+(1-b)pi(n) \qquad \forall j, j\neq i \qquad (4.2)$$

Here, a and b denote the reward and penalty associated with the automaton respectively.

As mentioned earlier, the learning automaton works perfectly in dynamically changing terrain surroundings in which the functional duration is directly proportional to the change in environment. There are many studies that clearly show that the LA-based methodologies are also best suited for solving optimization problems with a higher level of uncertainty. Since the coverage in WSNs is also one of the maximization problems where the objective is to prolong the functional duration, learning-automata-based metaheuristics certainly helps in achieving a prolonged network lifetime. The following section discusses some already working metaheuristics that are influenced by reinforcement learning while providing coverage and connectivity in the WSNs.

4.5.3.2 Connected Coverage Using Learning Automata

It was clearly mentioned during the detailed discussion in the last section that *LA* has been proven to be an important paradigm to solve various hard optimization-based problems in dynamically changing environments. Thus, the target coverage can be purposefully solved with learning-automaton-based approaches, which certainly suggest better solutions compared to other conventional heuristics in WSNs. Recently, many learning-automata-inspired paradigms have been addressed to prolong total network lifetime while providing connected coverage in the deployed networks [42–47].

Mostafaei et al. [43] addressed a learning-automaton-inspired scheduling algorithm for the target coverage problem to prolong the network's functional duration. Similarly, another *LA*-based protocol was addressed in [44] to maximize network lifetime for area coverage. Meybodi et al. [45] also presented another LA-inspired metaheuristic for solving the well-known target coverage issue in WSNs. Furthermore, Chand et al. [46] addressed an energy-efficient reinforcement metaheuristic with a learning-automata concept, where nodes with higher residual battery and maximum coverage of covered targets were preferred over the rest of the sensors in the deployed networks.

When comparing all these learning-automata-based metaheuristics, we can observe that most of the reinforcement methodologies basically comprise four major phases: the network setup phase, the learning phase, the monitoring phase, and the update phase. Most existing studies follow these phases in the same manner except the monitoring phase, where they need to decide which set of sensors are to be selected for connected coverage in order to prolong the total network lifetime. Now, we will give a brief overview of these major phases in order to understand the functionality of reinforcement learning-based metaheuristics.

Network setup: We assume that the network consists of N number of sensor nodes having a fixed sensing range R, which are randomly spread in a predefined area of size $M \times M$. In the same network, M targets are also deployed. Once the deployment process is over, these sensors and targets are fixed. Hence, these sensors in the network know which targets are covered by them individually. Every sensor Si has an associated LAi which helps sensor Si to determine the optimal action at any point in time in the given environment LE. As discussed in the previous section, a learning automaton of every sensor device has to choose between two actions at any point in time, namely: ACTIVE or INACTIVE. Further, a variable structure LA is opted for due to the changing dynamics of the deployed networks. At the time of the first iteration, both actions, be they ACTIVE or INACTIVE, have equal probability of 0.5. Once the network setup phase is over, we proceed to the next phase as follows.

Learning the environment: This is the backbone phase of any reinforcement learning-based metaheuristic. During this phase, each LAi (which corresponds to node Si) decides the future action to be taken by it, which can be either ACTIVE or INACTIVE. At first, each LA is assumed to be in an INACTIVE state. Every metaheuristic has its own criteria for calculating action probabilities based on numerous factors, and every LAi is modified such that sensor nodes with higher action probabilities are chosen in the current cover set and so they are in the ACTIVE state. Many metaheuristics give preference to those sensors that have higher remaining energy, and few of these paradigms choose sensors with higher coverage of uncovered targets. Chand et al. [46] somehow combined both parameters and claimed to have found the longest possible network lifetime compared to many existing studies on the same problem of solving connected coverage in the field of wireless sensor networks. Once LAi are updated as per the above selected criteria, then the next phase begins, where, based on the chosen action, the respective update phase is applied.

Monitoring: Once all these LAi are updated for these sensors Si, the next phase is to select sensor nodes whose LA has the highest action probability attached in order to cover all the targets and then connect all the sensors to the BS. As discussed, in the previous phase, the LA of each sensor is updated according to the parameter selected by the respective metaheuristics chosen. As we know, the deployed network becomes non-functional when the first target becomes uncovered. As we also observed in the previous chapter, the targets that are covered by a minimum of sensing devices, known as critical targets, become uncovered first. While prolonging the network lifetime, all the research studies [43–48] addressed many solution heuristics based on learning automata where the objective was to select those sensors which have either higher remaining energy, cover the maximum uncovered targets, or both. This is the only phase where all the studies differ in the way they choose sensors for the next iterations. Once the monitoring phase is over, the next task is to update the energy of the sensors active in the current cover set and also reward or penalize their *LA* to make them ready.

Update: Once cover sets are formed and connectivity with the base station is also ensured, the batteries of the sensors whose *LAs* are in an ACTIVE state in the current iteration should be updated. Further, their respective *LAs* should be either rewarded or penalized based on the threshold criteria set on the action probability vector of the networks. There are many methods to decide whether to reward or penalize the *LA* for the action taken.

Next, we present a comparative study of a few existing research studies in the literature [42, 43, 46, 48]. All these studies follow the four steps discussed in the previous section, namely setup phase, learning, monitoring, and update. While studying this existing research closely, we found that all the methodologies are similar except for the monitoring phase, which differs in how the metaheuristics select the sensors to provide coverage as well as connectivity in the deployed networks. Moreover, the update phase is also similar in all these paradigms. In the next section, we will give a brief description of the main four metaheuristics inspired by reinforcement learning in order to compare their outcomes [42, 43, 46, 48].

The overall functionality of any of the metaheuristics [42, 43, 46, 48] can be easily described with the help of a flow chart, as shown in Figure 4.13.

4.5.4 Simulation and Experimentation

In this section, we will discuss a few existing metaheuristics addressed on the connected coverage issue using learning-automata paradigms. In order to understand these methodologies properly, we will first give a brief description of the metaheuristics and then we will discuss some important observations conceived from the detailed experimentation.

4.5.4.1 Simulation Setup

For the simulation, we considered a heterogeneous network spread over a squared region of two dimensions, namely 200*200 m^2 and 300*300 m^2. All the sensors have a fixed and identical sensing range of 100 m. We also assume a randomly generated network where sensor and target X and Y coordinates are generated using a pseudorandom generation technique. The simulation is carried out on MATLAB (2016) on a Core i3 processor with a 2.10 GHz processor and 4 GB RAM.

4.5.4.2 Existing Metaheuristics

There have been many studies done on connected coverage using learning automata in WSNs so far. Here, we consider a few relevant metaheuristics for experimentation and simulation of their performance. Before going into a detailed performance

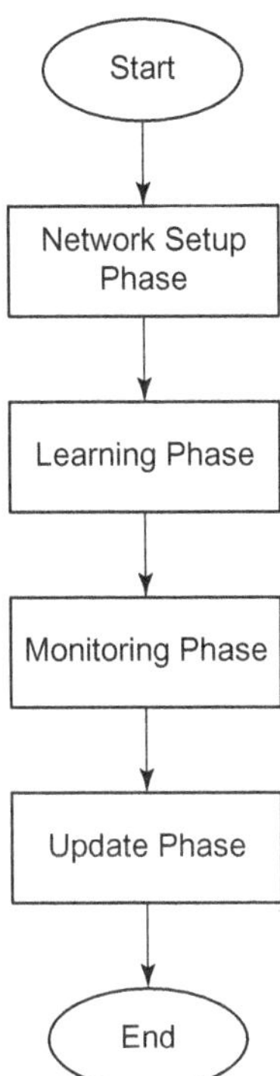

FIGURE 4.13 Flow chart of metaheuristics

comparison, we would like to briefly describe them showing how these metaheuristics work and their observed drawbacks.

Mohamadi et al. [42] address their first metaheuristic, which is inspired by the reinforcement learning paradigm, for the well-known target coverage problem. During the monitoring phase, our proposed heuristics primarily give importance to those sensors covering the maximum uncovered targets. Based on this, the *LAs* of each contributing sensor and action probability vector are updates. We considered a static network in most applications, where the coverage of an individual sensor does not change over time. Therefore, this methodology keeps selecting the same set of sensors in consecutive cover sets, resulting in early energy holes in the deployed networks. Therefore, the sensors participating in consecutive rounds become exhausted after a few hours of deployment, leading to early energy holes despite many sensors in the same network still having full energy reserves.

Mostafaei et al. [43] addressed another reinforcement learning-based metaheuristic where, during the monitoring phase, they selected sensors and then randomly chose their action (ACTIVE or INACTIVE), and they followed this process until a cover set was formed. Further, the chosen action was either rewarded or penalized based on the selected parameters by the metaheuristic proposed. In this study, if a sensor was covering an altogether uncovered target, then its action is rewarded; otherwise, it is penalized if the target is already covered by other neighbouring sensors. The study in [46] follows the two rules when monitoring targets for an extended duration in deployed sensor networks.

Rule 1: Here, identify all the critical targets and then also identify critical sensors (those that cover critical targets), and then select only that critical sensor with a higher ACTIVE action probability than INACTIVE and that has a higher energy ratio with maximum uncovered targets coverage. By doing this, authors ensured the longevity of the network lifetime, as critical sensors' functional duration directly impacts the network lifetime. Once this step is followed by the designed metaheuristic, the authors followed rule 2.

Rule 2: As we know, the deployed network is dense and targets are monitored by multiple nodes in close proximity. Thus, in each cover set, there is redundant coverage for many targets, which results in reduced network lifetime. In order to resolve this issue, the metaheuristic in [46] further optimized the generated cover set by minimizing the number of sensors in each cover set. Here we say that the minimal set is also a cover set, but this ensures that targets do not have redundant coverage.

Finally, the authors in work [48] proposed a thorough approach followed by a learning-automata-based heuristic where sensors with higher coverage are selected for forming the cover set, and there is a set of basic rules to either reward or penalize the action probability of the selected sensors.

4.5.4.3   Comparative Performance Analysis

Now, we discuss the performance comparisons between the above-discussed metaheuristics for the connected target coverage where the objective is to provide coverage and connectivity in WSNs. For experimentation, we consider the homogeneous sensor network. First, we observe the effect of varying nodes in a fixed area. To do so, we have chosen the metaheuristic used by the authors in [46], whose detailed functionality is explained in the previous section. For simulation purposes, we consider sensors between 100 and 300, and targets between 50 and 200 with a fixed sensing range of 200 m. For each value of the outcome, we have considered an average of 50 instances. Thus, depending on the simulation, we chose the number of sensors and targets accordingly.

4.5.4.4   Sensors and Targets vs. Network Functional Duration

Here, we will study how the network functional duration directly depends on the varying number of sensors and targets in the fixed networks while providing connected coverage. In order to understand the impact of varying sensors, consider Figure 4.14, where there are varying sensors in the range of 100 to 300 and fixed targets (50). As shown in Figure 4.14, we can observe that network lifetime increases with more sensor nodes. This is because with a greater number of sensors, the network will operate for an extended duration, which in turn extends the network lifetime overall.

Further, we experimented with varying targets between 50 and 250 and fixed sensors (100) in the same fixed-size network, as shown in Figure 4.15. Here, we observe that the network lifetime decreases with more targets deployed in the same area.

This happens due to the fact that more sensors are required to cover an increasing number of targets, which in turn results in a shorter network lifetime. So far, we have observed that when we increase sensors in a fixed network while maintaining connected coverage, the network's functional duration will increase. Conversely, when we deploy more targets, the functional duration will certainly decrease due to the extended connectivity of the monitoring sensors.

4.5.4.5 Network Functional Duration vs. Sensing Range

Now, we address the impact of the sensing range of the sensors on the functional duration of the network while maintaining the connected coverage. We experimented with 100 sensors and 20 targets, and here we consider sensing ranges varying between 100 and 500, as shown in Figure 4.16.

As depicted in the figure, we can observe that network functional duration extends in the sensing range of the sensors in the same fixed area while still providing connected coverage in the WSN. The major reason behind this increase is the extended coverage in the same close proximity by the same set of sensors, which result in a prolonged network functional duration.

Thus far, we have observed that the network functional duration is directly dependent on the deployed sensors, targets, and the sensing range assigned to these sensors while providing connected coverage in WSNs. Hence, we need to choose them wisely as per the application requirements.

Next, we experiment to compare the performance of the existing and discussed major reinforcement learning-based metaheuristics [42, 43, 46, 48] for providing connected coverage. For simulation purposes, we have compared their performance in terms of the network function duration.

4.5.4.6 Performance Comparison for Connected Coverage

Here, we experimented with performance comparisons of various existing learning-automata-based metaheuristics [42, 43, 46, 48] for providing connected coverage for

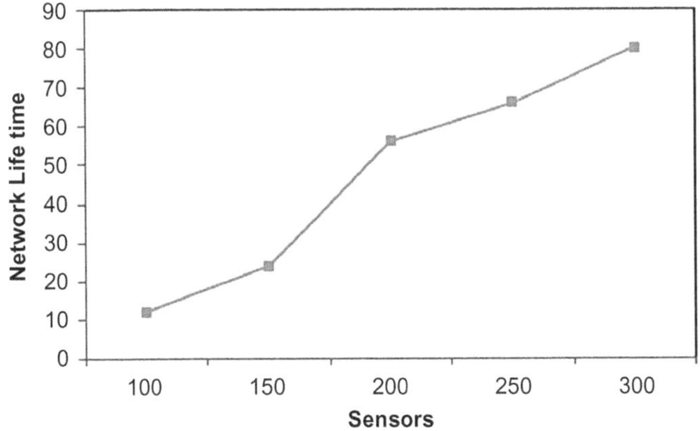

FIGURE 4.14 Network lifetime with varying sensors

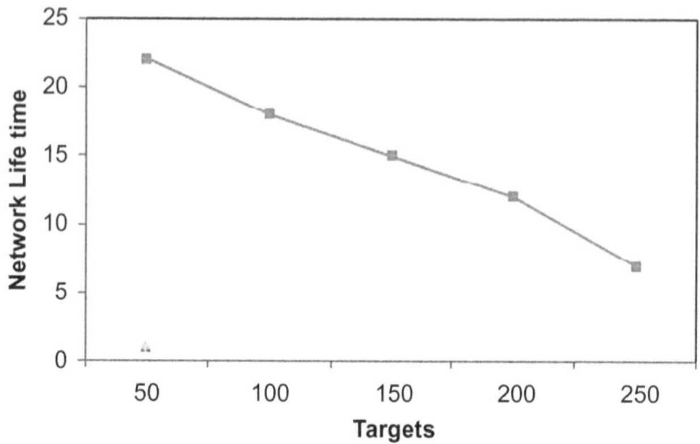

FIGURE 4.15   Network lifetime with varying targets

WSNs. For the simulation, we considered sensors varying between 50 and 250 and targets varying between 20 and 100. Further, we assume that the sensing range is fixed and equals 300 m within the sensing area of 200*200 m^2.

First, we compare the performance of these metaheuristics with varying sensors in the same fixed sensing area, with targets equal to 50 and a sensing range of 300 m. We take an average of 50 iterations to calculate a single value. As depicted in Figure 4.17, one can easily observe that the network lifetime achieved is directly proportional to the number of sensors deployed. Furthermore, various metaheuristics [42, 43, 46, 48] result in different outcomes based on their selected criteria for the monitoring phase of the learning-automata-based methodology.

The network functional duration is highest for the metaheuristic addressed in [46]. This is due to the fact that other metaheuristics [42, 43, 48] either consider the residual energy or the coverage of the uncovered targets of individual sensors. However, the metaheuristics proposed by the authors in [46] consider both

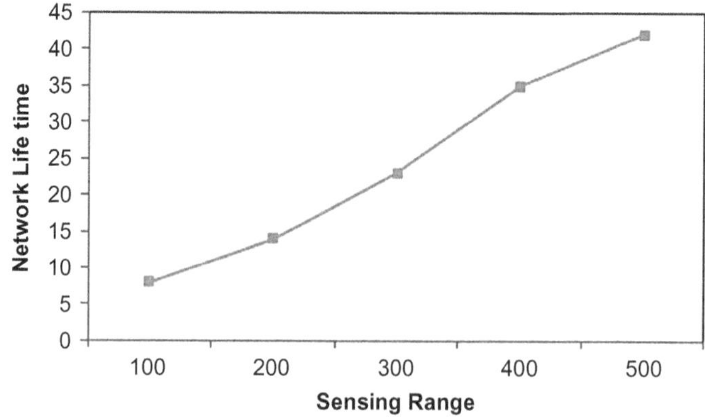

FIGURE 4.16   Network lifetime with varying sensing ranges

parameters together, which prolongs network functional duration compared to other studies. The same outcomes are clearly depicted in Figure 4.18.

Further, we simulated the same network configuration, only with one change: sensors are fixed to 100, and targets vary between 20 and 100. Here, again, we take an average of 50 iteration values to calculate a single value as shown in the graph in Figure 4.18. As depicted in Figure 4.18, we notice that the network lifetime achieved is greatly dependent on the targets deployed. With more targets, the network lifetime degrades, as more sensors are required to cover an extended number of targets. Furthermore, various metaheuristics [42, 43, 46, 48] result in different outcomes based on their selected criteria for the monitoring phase of the learning-automata-based methodology. The network functional duration is highest for the metaheuristic addressed in [46].

This happens because other metaheuristics [42, 43, 48] either consider the residual energy or the coverage of the uncovered targets of individual sensors. However, the metaheuristics proposed by the authors in [46] consider both parameters together, which increases network functional duration compared to other studies. These outcomes are clearly depicted in Figure 4.18.

4.6 CONCLUSION

In this chapter, we addressed connectivity and communication issues in WSNs. First, we addressed connectivity in WSNs and the associated issues of energy-optimized connectivity, where the major challenge is to efficiently utilize the limited battery associated with the sensing devices in the deployed networks. We addressed connected coverage where the objective is to monitor a certain number of targets and then ensure the selected set of sensors is also connected to the base station to forward the collected information. To provide connected coverage, we discussed many reinforcement learning-based metaheuristics, which are mainly based on learning

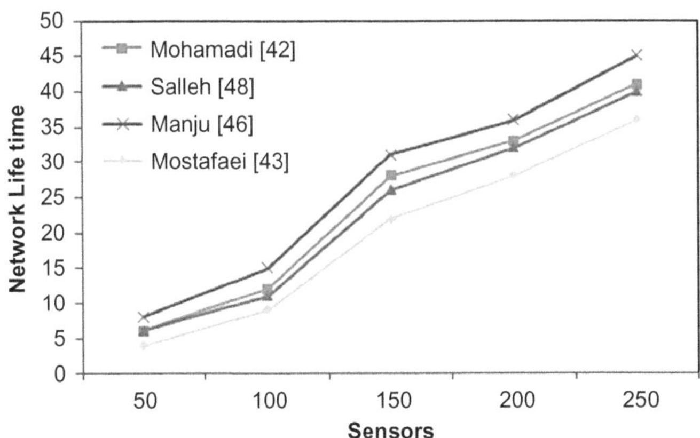

FIGURE 4.17 Performance comparison with varying sensors

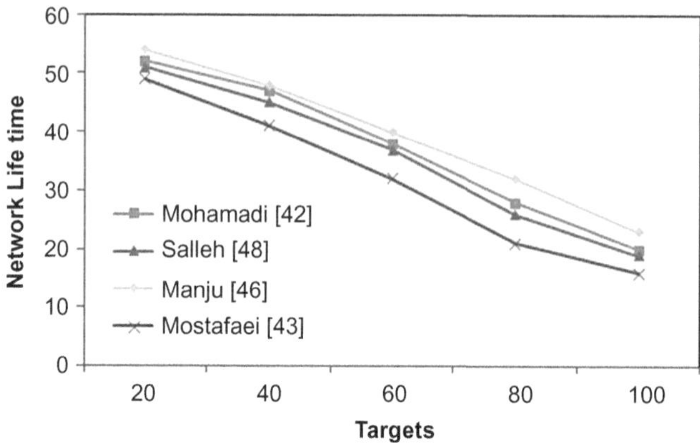

FIGURE 4.18　Performance comparison with varying targets

automata [42, 43, 46, 48]. We also performed a performance comparison with detailed simulation results. We experimented to check their performance of connected coverage with varying sensors, targets, and sensing ranges. In all the experiments, we observed that the work done by the authors in [46] outperformed the other metaheuristics as presented by the authors in [42, 43, 48].

REFERENCES

1. Akyildiz, I. F., Weilian Su, Y. Sankarasubramaniam, and E. Cayirci, "A Survey on Sensor Networks," *In IEEE Communications Magazine* 40, no. 8 (August 2002): 102–14. https://doi.org/10.1109/MCOM.2002.1024422.
2. Banimelhem, O., M. Naserllah, and A. Abu-Hantash, "An Efficient Coverage in Wireless Sensor Networks Using Fuzzy Logic-Based Control for the Mobile Node Movement." *2017 Advances in Wireless and Optical Communications (RTUWO), Riga, Latvia* (2017): 239–44. https://doi.org/10.1109/rtuwo.2017.8228541.
3. Farsi, M., M. A. Elhosseini, M. Badawy, H. Arafat Ali, and H. Zain Eldin. "Deployment Techniques in Wireless Sensor Networks, Coverage and Connectivity: A Survey." *In IEEE Access* 7 (2019): 28940–54.
4. Manju and Arun K. Pujari "Target Q-Coverage Problem in Wireless Sensor Networks." International Conference on Distributed Computing Networks (ICDCN'09), Hyderabad, January, pp. 325–30, 2009.
5. Ketshabetswe, L. K., A. M. Zungeru, M. Mangwala, J. M. Chuma, and B. Sigweni. "Communication Protocols for Wireless Sensor Networks: A Survey and Comparison." *Heliyon* 5, no. 5 (May 2019): e01591.
6. Manju, Deepti Singh, Satish Chand, and Bijendra Kumar. "Target Coverage Heuristic in Wireless Sensor Networks." 10th Int. Conf. CACCT 2016: Advanced Computing and Communication Technologies, Springer, pp. 265–73, 18h–20 November, 2016
7. Tripathi, A., H. P. Gupta, T. Dutta, R. Mishra, K. K. Shukla, and S. Jit. "Coverage and Connectivity in WSNs: A Survey, Research Issues and Challenges." *In IEEE Access* 6 (2018): 26971–92.
8. Dimple, Manju. "Energy-Efficient Algorithm for Target Coverage Problems in Wireless Sensor Networks." *ICACCT-16*, pp. 124–7, November, 2013, Panipat Haryana.

9. Xu, Yun, Wanguo Jiao, and Mengqiu Tian. "Energy-Efficient Connected-Coverage Scheme in Wireless Sensor Networks." *Sensors* 20, no. 21 (2020): 6127.

10. Sheikh-Hosseini, Mohsen, Seyed Rouhollah, and Samareh Hashemi. "Connectivity and Coverage Constrained Wireless Sensor Nodes Deployment Using Steepest Descent and Genetic Algorithms." *Expert Systems with Applications* 190 (2022). https://doi.org/10.1016/j.eswa.2021.116164.

11. Tirandazi, P., A. Rahiminasab, and M. J. Ebadi. "An Efficient Coverage and Connectivity Algorithm Based on Mobile Robots for Wireless Sensor Networks." *Journal of Ambient Intelligence and Humanized Computing* (2022). https://doi.org/10.1007/s12652-021-03597-9.

12. Manju, Arun K Pujari. "High-Energy-First (HEF) Heuristic for Energy Efficient Target Coverage Problem." *International Journal of Ad-Hoc Sensor and Ubiquitous Computing* 2, no. 1 (March 2011): 45–58.

13. Singh, Samayveer, Manju, Aruna Malik, and Pradeep Kumar Singh. "A Threshold-Based Energy Efficient Military Surveillance System Using Heterogeneous Wireless Sensor Networks." *Soft Computing* 27 (July 2021): 1–14.

14. Manju and Samayveer Singh. "Modified Energy-Proficient Partial Coverage Methodology for Optimizing Coverage in WSN." *Multimedia Tools and Applications* (2023). https://doi.org/10.1007/s11042-023-14563-2.

15. Ketshabetswe, Lucia Keleadile, Adamu Murtala Zungeru, Mmoloki Mangwala, Joseph M. Chuma, and Boyce Sigweni. "Communication Protocols for Wireless Sensor Networks: A Survey and Comparison." *Heliyon* 5, no. 5 (2019). https://doi.org/10.1016/j.heliyon.2019.e01591.

16. Ngadi, A., and V. C. Gungor. "Smart Grid Communication and Information Technologies in the Perspective of Industry 4.0 : Opportunities and Challenges." *Computer Science Review* 30 (2018): 1–30.

17. Manju, Samayveer Singh, Sandeep Kumar, Anand Nayyar, Fadi Al-Turjman, and Leonardo Mostarda. "Proficient QoS-Based Target Coverage Problem in Wireless Sensor Network." *IEEE Access* 8 (2020). https://doi.org/10.1109/access.2020.2986493.

18. Jang, I. et al. "A Survey on Communication Protocols for Wireless Sensor Networks." *Journal of Computing Science and Engineering. Korean Institute of Information Scientists and Engineers* (2013). https://doi.org/10.5626/jcse.2013.7.4.231.

19. Kochhar Aarti, Pardeep Kaur, Preeti Singh, and Sukesha Sharma. "Protocols for Wireless Sensor Networks: A Survey." *Journal of Telecommunications and Information Technology* 1, no. 1 (April 2018): 77–87.

20. Feng, Xiang, Fang Yan, and Xiaoyu Liu. "Study of Wireless Communication Technologies on Internet of Things for Precision Agriculture." *Wireless Personal Communications* 108 (2019): 1785–802.

21. Kim, Beom-Su, Ho Sung Park, Kyong Hoon Kim, Daniel Godfrey, and Ki-Il Kim. "A Survey on Real-Time Communications in Wireless Sensor Networks." (2017).

22. Landaluce, H., L. Arjona, A. Perallos, F. Falcone, I. Angulo, and F. Muralter. "A Review of IoT Sensing Applications and Challenges Using RFID and Wireless Sensor Networks." *Sensor* 20, no. 9 (2020): 2495.

23. Ahlawat, Meena, and Ankita Mittal. "Different Communication Protocols for Wireless Sensor Networks: A Review." *International Journal of Advanced Research in Computer and Communication Engineering* 4, no. 3 (March 2015). https://doi.org/10.17148/ijarcce.2015.4351.

24. Hortelano, D., T. Olivares, M. C. Ruiz, C. Garrido-Hidalgo, and V. López. "From Sensor Networks to Internet of Things. Bluetooth Low Energy, a Standard for This Evolution." *Sensors (Basel)* 17, no. 2 (2017): 372.

25. Hughes, Josie, Jize Yan, and Kenichi Soga. "Development of Wireless Sensor Network Using Bluetooth Low Energy (BLE) for Construction Noise Monitoring." *International Journal on Smart Sensing and Intelligent Systems* 8, no. 2 (2015): 1379–405.
26. Ijemaru, Gerald K., Kenneth Li-Minn Ang, and Jasmine K. P. Seng. "Wireless Power Transfer and Energy Harvesting in Distributed Sensor Networks: Survey, Opportunities, and Challenges." *International Journal of Distributed Sensor Networks* 18, no. 3 (2022). https://doi.org/10.1177/15501477211067740.
27. Ouni, Ridha, and Kashif Saleem. "Framework for Sustainable Wireless Sensor Network Based Environmental Monitoring." *Sustainability* 14, no. 14 (2022): 8356.
28. Dash, Biswajit Kumar, and Jun Peng. "Zigbee Wireless Sensor Networks: Performance Study in an Apartment-Based Indoor Environment." *Hindawi Journal of Computer Networks and Communications* 2022 (2022): 14.
29. Tuan Tran, M. A., T. N. Le, and T. P. Vo. "Smart-Config Wifi Technology Using ESP8266 for Low-Cost Wireless Sensor Networks." 2018 International Conference on Advanced Computing and Applications (ACOMP), Ho Chi Minh City, Vietnam, 2018, pp. 22–8, 2018.
30. Samijayani, N., R. Darwis, S. Rahmatia, A. Mujadin, and D. Astharini. "Hybrid ZigBee and WiFi Wireless Sensor Networks for Hydroponic Monitoring." 2020 International Conference on Electrical, Communication, and Computer Engineering (ICECCE), Istanbul, Turkey, pp. 1–4, 2020.
31. Raza, U., P. Kulkarni, and M. Sooriyabandara. "Low Power Wide Area Networks: An Overview." *In IEEE Communications Surveys & Tutorials* 19, no. 2 (2017): 855–73.
32. Singh, Samayveer, Anshu Kumar Dwivedi, A. K. Sharma, and Pawan Singh Mehra. "Learning Automata Based Heuristics for Target Q-Coverage." *ICRITO* (January 2020). https://doi.org/10.1109/icrito48877.2020.9197923.
33. Akram, Junaid, Hafiz Suliman Munawar, Abbas Z. Kouzani, and M. A. Parvez Mahmud. "Using Adaptive Sensors for Optimised Target Coverage in Wireless Sensor Networks." *Sensors* 3 (2022): 1083.
34. Banoth, S. P. R., P. K. Donta, and T. Amgoth. "Target-Aware Distributed Coverage and Connectivity Algorithm for Wireless Sensor Networks." *Wireless Network* 29 (2023): 1815–1830.
35. Elhabyan, Riham, Wei Shi, and Marc St-Hilaire. "Coverage Protocols for Wireless Sensor Networks: Review and Future Directions." *Journal of Communications and Networks* 21, no. 1 (2019): 45–60.
36. Yau, K. L. A., H. G. Goh, D. Chieng, et al. "Application of Reinforcement Learning to Wireless Sensor Networks: Models and Algorithms." *Computing* 97 (2015): 1045–75.
37. Mostafaei, Habib, Antonio Montieri, Valerio Persico, and Antonio Pescapé. "A Sleep Scheduling Approach Based on Learning Automata for WSN Partial Coverage." *Journal of Network and Computer Applications* 80 (2017): 67–78.
38. Harizan, Subash, and Pratyay Kuila. "Evolutionary Algorithms for Coverage and Connectivity Problems in Wireless Sensor Networks: A Study." *Design Frameworks for Wireless Networks* (2020): 257–80. https://doi.org/10.1007/978-981-13-9574-1_11.
39. Le Nguyen, Phi, et al. "Node Placement for Connected Target Coverage in Wireless Sensor Networks with Dynamic Sinks." *Pervasive and Mobile Computing* 59 (2019): 101070.
40. Zhang, Zhen, Dongqing Wang, and Junwei Gao. "Learning Automata-Based Multiagent Reinforcement Learning for Optimization of Cooperative Tasks." *IEEE Transactions on Neural Networks and Learning Systems* 32, no. 10 (October 2021). https://doi.org/10.1109/tnnls.2020.3025711.
41. Narendra, K. S., M. A. L. Thathachar. *Learning Automata: An Introduction.* Englewood Cliffs: Prentice-Hall, 1989.

42. Mohamadi, H., A. S. Ismail, and S. Salleh. "Solving Target Coverage Problem Using Cover Sets in Wireless Sensor Networks Based on Learning Automata." *Wireless Personal Communications* 75 (2014): 447–63.
43. Mostafaei, H., and M. R. Meybodi. "Maximizing Lifetime of Target Coverage in Wireless Sensor Networks Using Learning Automata." *Wireless Personal Communications* 71, no. 2 (2013): 1461–77.
44. Mostafaei, H., M. Meybodi, and M. Esnaashari. "A Learning Automata-Based Area Coverage Algorithm for Wireless Sensor Networks." *Journal of Electronic Science and Technology* 8, no. 3 (2010): 200–5.
45. Mostafaei, H., M. Esnaashari, and M. R. Meybodi. "A Coverage Monitoring Algorithm Based on Learning Automata for Wireless Sensor Networks." *Applied Mathematics & Information Sciences* 9, no. 3 (2015): 1317–25.
46. Manju, Satish Chand, and Bijendra Kumar. "Target Coverage Heuristic Based on Learning Automata in Wireless Sensor Networks." *IET Wireless Sensor System* (2017). https://doi.org/10.1049/iet-wss.2017.0090.
47. Manju, Samayveer Singh, Anshu Kumar Dwivedi, A. K. Sharma, and Pawan Singh Mehra. "Learning Automata Based Heuristics for Target Q-Coverage." 2020 8th International Conference on Reliability, Infocom Technologies and Optimization (Trends and Future Directions) (ICRITO), pp. 170–73, 2020.
48. Mohamadi, H., A. Ismail, S. Salleh, et al. "Learning Automata-Based Algorithms for Finding Cover Sets in Wireless Sensor Networks." *Journal of Supercomputing* (in press) 66, no. 3 (2013): 1533–52.

Chapter 5

SUMMARY

Chapter 5 presents a concise study of energy-efficient clustering in wireless sensor networks (WSNs), highlighting their significance in enhancing network longevity and efficacy. The chapter simplifies the concept of clustering, where sensor nodes are grouped and managed by cluster heads (CHs) to improve communication and reduce energy use. The chapter evaluates both metaheuristic algorithms and traditional methods for CH selection, balancing advanced techniques with the need for simplicity in certain network environments.

The narrative addresses the "energy hole" issue – areas of rapid energy depletion that threaten network stability – and proposes clustering as a remedial strategy. It assesses the problem's impact and outlines adaptive measures for effective energy management.

Moreover, the text discusses data dissemination, showing how clustering aids in efficient data flow and energy conservation. Performance analysis underscores network lifetime, load balancing, and throughput, with an aim to refine network operations and energy utilization.

In summary, the chapter underscores the vital role of strategic clustering in WSNs, advocating for the careful selection of algorithms to combat energy holes and achieve sustainable network performance. It offers an integrated view of theoretical and practical approaches to energy-efficient clustering for WSN optimization.

DOI: 10.1201/9781003427780-9

5 Energy-Efficient Clustering in Sensor Networks

5.1 INTRODUCTION TO ENERGY-EFFICIENT CLUSTERING

In today's digitally driven society, the reliance on advanced technologies is escalating, pushing the boundaries of energy consumption to new heights. The digital landscape, marked by the proliferation of wireless sensor networks (WSNs), the Internet of Things (IoT), and expansive cloud computing capabilities, is at the forefront of technological innovation. However, this progress comes at a cost, notably in the form of increased energy demands, which pose significant environmental and economic challenges. Over recent years, the potential of WSN technology has captured the attention of both the academic community and industry leaders, given its wide-ranging applications across various sectors [1]. WSNs find their utility in several key areas including, but not limited to, industrial processes, home automation, healthcare monitoring, and defence applications [2]. Comprising numerous small, distributed nodes, WSNs facilitate the collection and transmission of data across vast areas, employing either wired or wireless communication to establish a network [3]. These sensor nodes are strategically deployed across targeted zones to monitor environmental conditions, aggregating and transmitting the collected data wirelessly to a central node, known as the sink or base station (BS), for further analysis. Beyond environmental monitoring, these nodes are also capable of acting as relays or data fusion centres, routing information from adjacent nodes to the BS [4]. The BS then processes this data locally or serves as a conduit to relay information to remote servers for additional analysis. The versatility and efficient data distribution capabilities of WSNs underpin their broad applicability, highlighting their role in fostering adaptable infrastructural solutions across diverse fields [5]. The major applications of WSNs are shown in Figure 5.1.

Sensor nodes are equipped with a broad array of sensors tailored for various monitoring tasks, such as pressure, sound, light, weather conditions, temperature, and chemical composition. These nodes are integral components of a sensor network, as illustrated in a typical diagram (refer to Figure 5.2). A significant challenge facing these networks is the limited lifespan of their batteries, which introduces two main hurdles in maintaining operational efficiency [1]. First, the logistical and financial burden of regularly servicing a vast number of nodes scattered over extensive areas

DOI: 10.1201/9781003427780-10

FIGURE 5.1   Application areas of WSNs

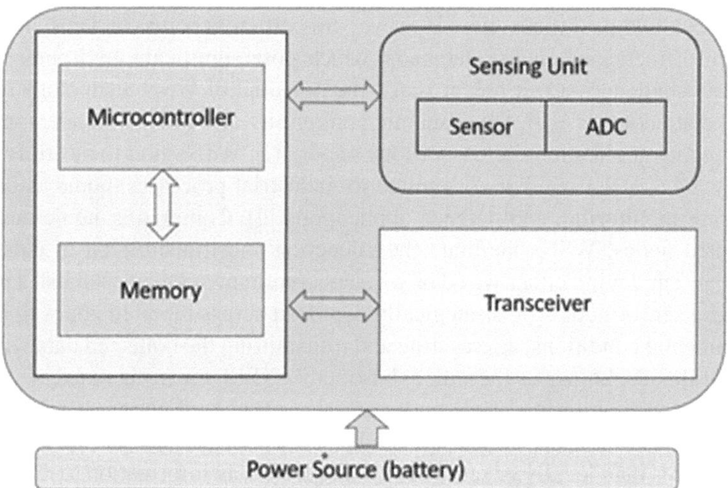

FIGURE 5.2   Structure of a WSN node

is considerable. Second, accessing these nodes for battery replacement or charging can be impractical or impossible due to their locations. This is especially true for nodes embedded within structures, situated in hazardous environments, or deployed in areas of strategic military importance, where physical retrieval or maintenance poses significant risks [2].

Clustering within WSNs is a critical strategy for enhancing energy efficiency and extending the overall life of the network. This approach involves grouping sensor nodes into clusters, with one node acting as the cluster head. This cluster head is responsible for direct communication with the base station. By implementing data aggregation at this level and reducing the necessity for long-range transmissions, the system significantly conserves energy. Algorithms designed for clustering play a vital role in distributing energy consumption evenly among nodes, preventing network bottlenecks, and improving the scalability of the network. Additionally, these algorithms support effective data fusion and aggregation, ensuring that data processing and decision-making at the base station are both efficient and informed.

The significance of energy-efficient clustering in WSNs cannot be overstated, given its profound effect on the network's durability, reliability, and operational efficacy. Through the strategic grouping of sensor nodes into clusters and the appointment of cluster heads for the management of data collection and communication, this methodology drastically reduces the energy drain associated with transmitting information over long distances. This not only preserves the battery life of the nodes but also prolongs the network's functional duration. Furthermore, energy-efficient clustering methods promote equitable energy usage across the network, averting early exhaustion of individual nodes. They also bolster the network's ability to scale and maintain resilience by optimizing the use of available resources and minimizing congestion. Ultimately, the practice of energy-efficient clustering proves to be essential for optimizing the functionality, dependability, and sustainability of WSNs across a broad spectrum of uses, from monitoring environmental changes to streamlining industrial processes.

5.1.1 Major Challenges in WSNs

WSNs encounter a variety of hurdles that significantly influence their conception, deployment, and functionality. These issues include:

Limited energy resources: Given that sensor nodes are typically battery-operated, managing their limited energy to extend the network's lifespan is crucial.

Node placement: Strategically positioning sensor nodes to ensure comprehensive area coverage without overlaps or gaps presents a logistical challenge.

Network expandability: The ability of a WSN to scale efficiently, managing a larger network's data flow and interactions without depleting resources, is a key concern.

Data processing and integration: Implementing methods to aggregate and fuse data, reducing redundancy and optimizing transmission to save energy, is necessary.

Security measures: Protecting the network against security threats such as data breaches, node tampering, and service disruptions in a resource-limited setting is complex.

Service quality: Meeting specific quality of service benchmarks such as speed, dependability, and data throughput, while balancing energy conservation, involves significant compromise.

Accurate node localization: Determining precise locations for nodes in a cost-effective way, which is critical for many applications, remains a difficult task.

Adaptability to environmental shifts: WSNs deployed in fluctuating conditions must reliably adapt and maintain data integrity amid these changes.

Data handling: Efficiently managing the vast data generated by sensor nodes, from storage to retrieval, is crucial for meaningful analysis.

Autonomous network adjustment: In dynamic network environments, enabling nodes to self-adjust and maintain optimal operation autonomously is challenging.

Efficient data transmission: Crafting an effective strategy for data routing that conserves energy while ensuring data reaches its destination despite potential node failures or weak links is essential.

5.2 SIGNIFICANCE OF CLUSTERING IN WSNS

WSNs are inherently dynamic and form ad hoc networks, where managing their network topology emerges as a crucial issue. This management encompasses a broad range of aspects including resource allocation, network scalability, reliability, and operational efficiency [6]. Topology management stands out as a strategic approach to maintain the robustness, dependability, and effectiveness of ad hoc networks like WSNs [7]. Within this framework, identifying potential neighbouring nodes to establish connections and selecting the optimal neighbours for efficient data relay are pivotal. These decisions significantly contribute to improving network scalability, maximizing resource efficiency, ensuring network reliability, and addressing other pertinent challenges. Employing clustering and routing strategies serves multiple purposes: it aids in distributing work evenly across the network, avoiding data traffic collisions, consolidating data to reduce transmission loads, optimizing coverage and connectivity, reducing energy expenditure, decreasing data delivery times, extending the network's operational life, enhancing the network's ability to scale, and bolstering fault tolerance [8].

5.2.1 CLUSTERING PROCESS

Clustering in the context of WSNs is a strategic topology management technique aimed at enhancing network efficiency. This method organizes nodes into groups, distributing responsibilities among them to promote equitable resource usage and duty sharing. Each cluster comprises a set of member nodes overseen by one or more CHs. These CHs play a critical role in coordinating the activities within the cluster, including data collection, aggregation, processing, and transmission. Additionally, the network architecture typically includes one or more BSs, which function either as central data collection points or as gateways facilitating local data processing. Communication between CHs and BSs can occur directly or indirectly, with intermediary nodes facilitating the transmission over longer distances. However, the extensive energy demands for sustaining such a network architecture, especially due to the high energy consumption associated with long-distance transmissions, pose a significant challenge. Direct communication between sensor nodes and the BS, or even multi-hop routing involving several nodes, is not a viable solution due to the rapid depletion of node energy reserves. This limitation underscores the inherent challenge in WSNs related to energy conservation and the difficulty of maintaining long-distance communications without compromising network longevity and performance. Figure 5.3 illustrates a simplistic communication model within WSN, focusing on a single-layer communication framework.

The high energy demand of direct node-to-node communication is a notable limitation, often leading to redundancy in data from proximate sensors and premature depletion of distant nodes. To mitigate these issues, a hierarchical, two-tier approach organizes nodes into clusters, as illustrated in Figure 5.4. In this setup, a designated cluster head is responsible for aggregating and forwarding data to the base station, creating an efficient data flow structure with two distinct layers: one for the CHs and another for the subordinate nodes. These nodes regularly forward their data to the CH, which then consolidates and transmits it to the BS. However, this method places a heavier energy burden on the CH, especially for long-distance transmissions, potentially exhausting its energy supply prematurely [9]. To prevent this, the role of the CH is rotated among nodes to distribute energy consumption more evenly, ensuring the network's longevity and efficiency [9].

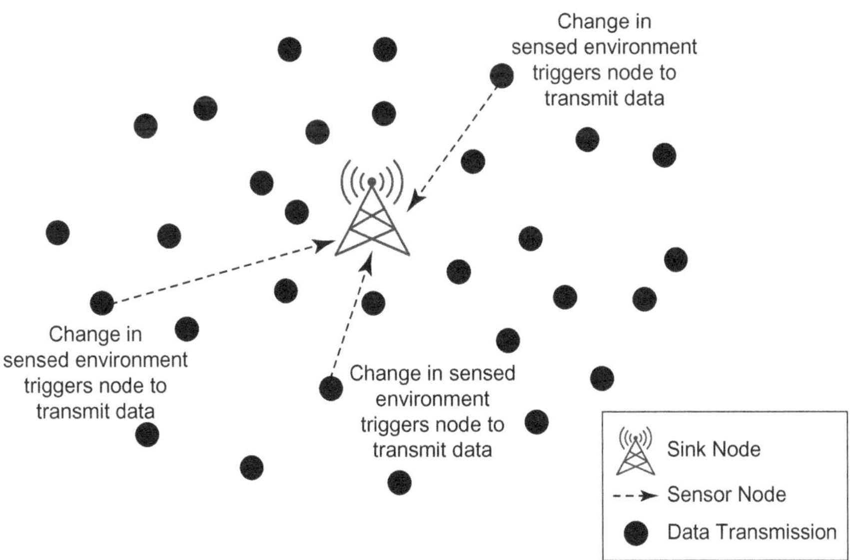

FIGURE 5.3   One-tier communication in a WSN

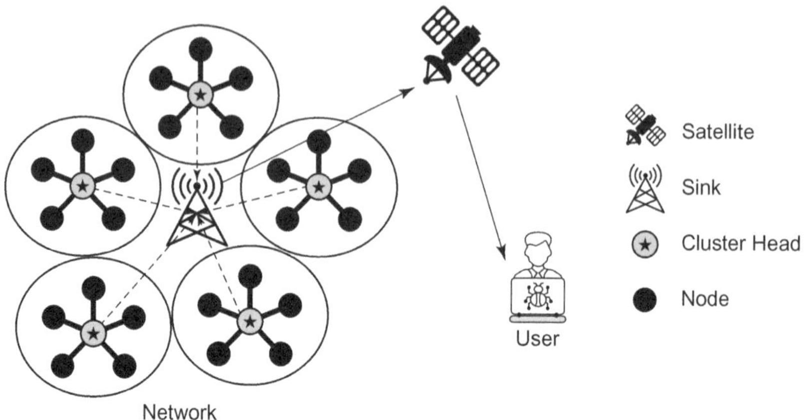

FIGURE 5.4 Hierarchical communication using clustering

5.2.2 Types of Clustering

Within the realm of WSNs, the strategic grouping of sensor nodes through clustering algorithms is pivotal for optimizing network functionality. These algorithms not only enhance energy conservation but also streamline communication processes and extend the overall lifespan of the network. Clustering algorithms are broadly divided into two main categories, reflecting their operational principles and optimization techniques: metaheuristic and non-metaheuristic approaches, as depicted in Figure 5.5.

Metaheuristic clustering algorithms focus on employing sophisticated search strategies to find near-optimal solutions for complex problems, where traditional methods may falter due to the vastness of the solution space or the intricate nature of the problem. These algorithms are inspired by natural phenomena and often incorporate mechanisms for exploration and exploitation to balance the search between globally optimal and locally optimal solutions. Examples include genetic algorithms, particle swarm optimization, and ant colony optimization, each drawing inspiration from biological processes and collective behaviour patterns to efficiently organize sensor nodes into clusters. Non-metaheuristic clustering algorithms, on the other hand, do not rely on nature-inspired search strategies but instead utilize deterministic or straightforward probabilistic methods to achieve clustering. These algorithms are typically faster and less computationally intensive than their metaheuristic counterparts, making them suitable for scenarios where quick decision-making is crucial. Common examples include the lowest ID algorithm, where the node with the lowest identifier within a vicinity becomes the cluster head, and the distance-based clustering algorithm, which selects cluster heads based on proximity to neighbouring nodes or a central point.

Both categories aim to optimize network performance by reducing energy consumption, minimizing communication distances, and balancing load among nodes.

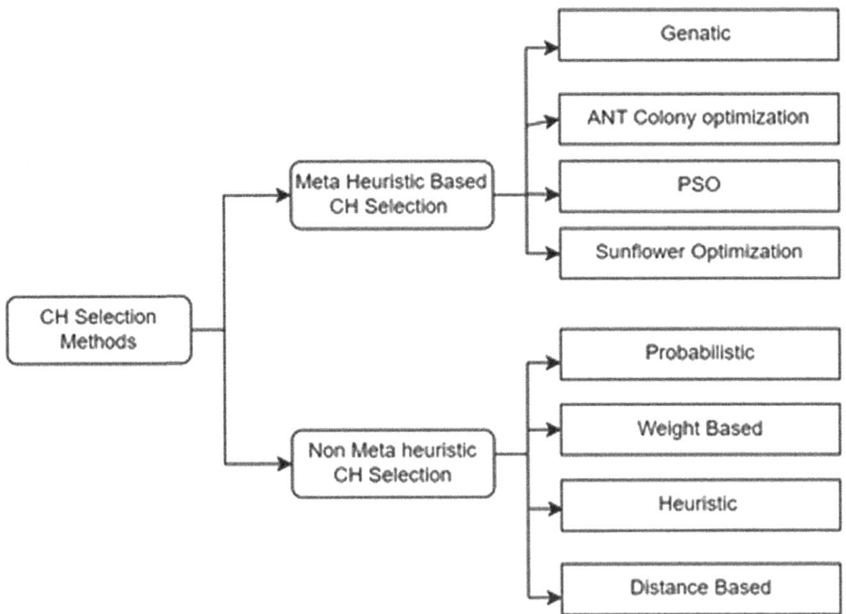

FIGURE 5.5   CH selection methods

By categorizing sensor nodes into clusters, these algorithms facilitate efficient data aggregation, minimize redundant transmissions, and, thus, significantly contribute to the prolongation of the network's operational lifetime. The choice between metaheuristic and non-metaheuristic approaches depends on the specific requirements and constraints of the WSN, including network size, node density, energy resources, and the desired balance between computational complexity and clustering effectiveness.

5.3   METAHEURISTIC-BASED CH SELECTION ALGORITHMS

A heuristic approach tailors strategies using specific problem insights to identify viable solutions. Conversely, metaheuristic algorithms are versatile, designed to be problem-agnostic, enabling their application across various domains with minimal adjustments needed for their inputs. This shift from a problem-specific to a problem-independent framework allows for the broad applicability of metaheuristic algorithms in solving diverse issues. Notable examples include genetic algorithms, bee colony optimization, particle swarm optimization, sunflower optimization, and ant colony optimization. These methods have been prominently applied in enhancing energy efficiency through clustering in WSNs, as Figure 5.6 illustrates, showcasing the primary metaheuristic techniques adopted by researchers in this field.

   The genetic algorithm for energy-efficient clustering in wireless sensor networks (GAECH) introduces a method that employs genetic algorithms for equitable energy

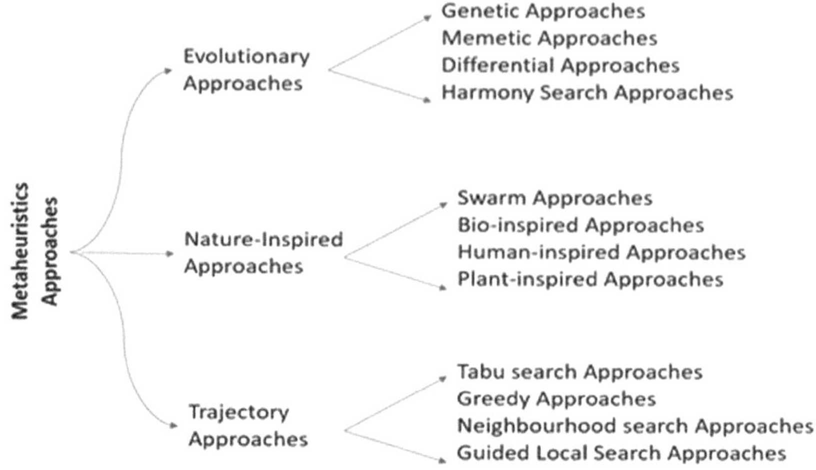

FIGURE 5.6   Types of metaheuristic approaches

distribution among cluster head (CH) nodes and for selecting the optimal CH from the sensor nodes [10]. This approach is guided by a multi-parameter fitness function, with its primary goal being the minimization of total energy consumption across both intra-cluster and inter-cluster communications. Consequently, the system ensures the formation of energy-efficient clusters. The fitness function incorporates a standard deviation measure to assess and minimize energy consumption disparities across clusters, facilitating balanced energy use. Additionally, CH-dispersion, another fitness function parameter, aims to optimize the spacing between CH nodes, thereby lowering energy expenditures in cluster-to-cluster communication. The fitness function also evaluates the workload on CH nodes, optimizing it to reduce the overall strain on these nodes.

In a different study, a mutation strategy for IoT-enabled heterogeneous WSNs in urban settings combines genetic algorithms with a greedy method, focusing on a weighted fitness function that considers node density, energy levels, and distances, alongside a three-tier network heterogeneity to prolong network lifetime and optimize energy deployment strategies [11]. This hybrid approach outperforms existing protocols by enhancing network longevity and selecting CHs based on comprehensive fitness evaluations.

Another enhancement targets the MGEAR protocol for heterogeneous WSNs, aiming to extend network lifespan and data transmission rates through a dual strategy of direct and multi-hop transmissions. This method ensures equitable energy distribution across nodes, considering their initial energy and proximity to the base station, and employs a central gateway to aggregate and forward data efficiently, demonstrating significant improvements in network durability and throughput [12]. Raslan et al. present an advanced sunflower optimization (ISFO) algorithm for CH selection in IoT contexts within WSNs. By integrating the Lévy flight operator, ISFO enhances the standard sunflower optimization by balancing exploration and exploitation, avoiding local optimum traps. Compared to six other swarm intelligence algorithms, ISFO shows superior performance in reducing energy consumption and extending network lifetime, establishing its efficacy in IoT WSNs [13]. Finally, Sangathir et al. introduce a hybrid algorithm combining the artificial bee colony and firefly algorithm (HMABCFA) for CH selection in WSNs. This innovative approach merges the advantages of both algorithms to address common optimization issues such as premature and slow convergence. The simulation outcomes reveal significant enhancements in network lifetime, stability, and latency, underscoring HMABCFA's role in refining WSN performance and energy efficiency [14].

5.4 NON-METAHEURISTIC-BASED CLUSTER HEAD SELECTION

This approach focuses on reducing energy consumption by randomly selecting cluster head nodes, overlooking crucial CH qualifications such as sufficient residual energy. Consequently, nodes with low energy might become CHs, neglecting the importance of strategic CH placement for efficient communication. Random selection can result in CHs being distant from both member nodes and the base station, thereby increasing communication costs within and between clusters. Figure 5.7 illustrates leading metaheuristic strategies researchers have employed for energy-efficient clustering in WSNs.

The low energy adaptive clustering hierarchy (LEACH) protocol [15] introduces a randomized approach for selecting cluster heads during its setup phase. Nodes not previously chosen as CHs generate a random number; if this number falls below a preset threshold, the node becomes a CH and broadcasts its status. Nodes with numbers above the threshold join the nearest CH as members. An advanced version, LEACH-C [16], enhances this process by having nodes communicate their energy levels and locations to the base station, which then selects the most suitable CHs, optimizing the selection process but raising concerns over scalability in larger WSNs.

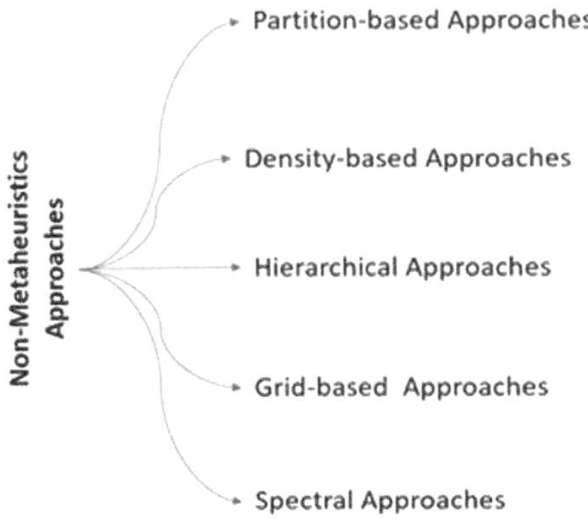

FIGURE 5.7 Types of non-metaheuristic approaches

Deterministic approaches for CH selection prioritize nodes with high residual energy and strategic locations to improve the energy efficiency and operational quality of clusters, aiming to prolong the WSN lifespan. However, these methods involve more complex cluster formation and CH selection processes, demanding additional computational and communication resources. The energy-efficient clustering algorithm based on neighbours (EECABN) [17] centralizes clustering by having nodes report their energy and location to the BS, which then chooses CHs based on a weighted calculation. The hybrid energy-efficient multipath routing (HEEMP) [18] strategy merges the advantages of flat and hierarchical routing, addressing the energy costs of distant member nodes by employing multi-hop transmissions within clusters and selecting energy-efficient paths for sending data to the BS.

Tyagi et al. [19] propose a novel threshold-based criterion for CH selection, focusing on a node's ability to serve as a CH without exhausting its energy. This involves estimating the number of rounds a node can function as a CH by dividing its remaining energy by its consumption in a communication cycle. As CHs near the end of their energy, they nominate successors considering energy levels and significance to the network, ensuring operational continuity. The model suggests a novel cluster-splitting method post-CH rotation to prevent energy depletion, wherein a CH divides its cluster by appointing two new CHs from its members, based on their proximity and energy status, thus balancing the load and enhancing network efficiency.

5.5 ROLE OF CLUSTERING IN ENERGY HOLE PROBLEMS

WSNs have surged in popularity over the last few decades, heralding a revolution in network technology that eschews the need for extensive physical infrastructure. These networks have found applications across a broad spectrum of fields, including but not limited to precision agriculture, domestic surveillance, environmental monitoring, varied military operations, and healthcare sectors [1, 2]. A typical sensor node within a WSN is outfitted with a limited supply of battery power and constrained memory capabilities. These nodes are complex assemblies that include a receiver, memory unit, sensor modules, a transmitter, and a power source. The sensor components are tasked with capturing environmental metrics, which may range from temperature, radiation levels, moisture content, to pressure variations. The collected data [3, 4] is then digitized and processed by an onboard microcontroller, before being relayed by the transmitter to a central base station, which facilitates data exchange over the internet regarding the monitored area.

Given the finite energy reserves of sensor nodes, optimizing energy consumption within WSNs is paramount. Clustering techniques have been widely adopted in numerous studies to enhance energy efficiency within these networks [20]. In this arrangement, sensor nodes are grouped into clusters, with one node within each cluster designated as the cluster head [9]. This CH node is responsible for aggregating environmental data from its cluster's sensors and forwarding this consolidated information to the BS [11]. While individual sensor nodes are capable of transmitting data directly to the BS, such transmissions, especially over long distances, are energy-intensive. Clustering mitigates this by enabling collective data forwarding

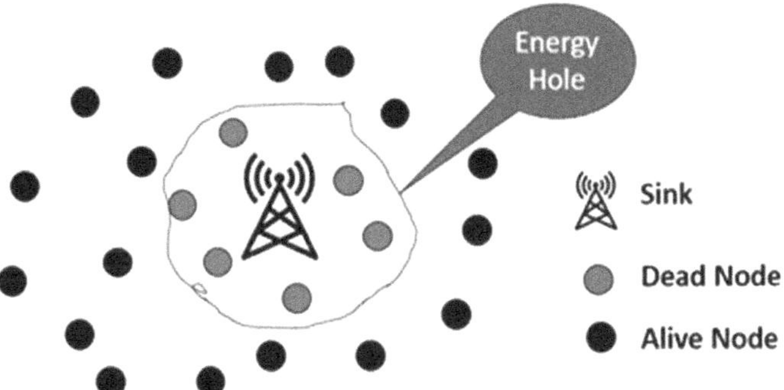

FIGURE 5.8 Problem of energy holes in the WSN

through the CHs, significantly conserving energy. However, CHs positioned further from the BS encounter increased energy depletion due to the longer transmission distances, leading to the formation of "energy holes" around the network, as shown in Figure 5.8 [21].

This phenomenon sees sensor nodes near the BS prematurely exhausting their energy reserves as they bear the brunt of data transmission, leading to operational failures despite other nodes retaining adequate energy levels. This uneven energy distribution and consumption necessitate strategic load management among CHs to sustain network efficiency and longevity.

5.5.1 Key Characteristics of the Energy Hole Problem in WSNs

The energy hole problem, inherently linked to the non-uniform energy depletion within the network, poses a grave threat to the operational sustainability of WSNs. It manifests most prominently in scenarios where closer nodes to the data collection point, or sink, deplete their energy resources at a faster rate than their more distant counterparts, leading to a scenario where data from parts of the network can no longer be relayed.

Uneven energy consumption: In WSNs, the issue of unequal energy consumption stands out prominently, particularly for nodes stationed near the central data collection point, or sink. These nodes bear the brunt of both communication duties and the task of relaying data from further nodes, leading to their accelerated energy drain in comparison to their more distantly located counterparts. This disproportionate energy usage not only affects the individual nodes but also the network's overall balance and efficiency.

Network partitioning risk: One significant repercussion of this uneven energy distribution is the heightened risk of network partitioning. As the nodes closest to the sink deplete their energy reserves and fail, they create what is known as an "energy hole." This phenomenon can sever connectivity paths within the network, causing certain segments to become isolated from the main data collection point. Such partitioning disrupts the flow of information, compromising the network's functionality and its ability to perform its intended tasks.

Reduced network lifetime: The presence of energy holes can substantially diminish the network's lifespan. The premature exhaustion of nodes within these critical areas impedes the network's capability for sustained monitoring and consistent data gathering. As these strategically important nodes cease to function, the network's overall operational efficiency and longevity are compromised, necessitating early interventions and potentially leading to increased maintenance and replacement costs.

5.5.2 Energy Hole Mitigation Techniques in WSNs

Research in the field of WSNs has been prolific, with scholars proposing various strategies to tackle the persistent challenge of energy hole formation, a problem that leads to uneven energy consumption and potentially shortens the network's lifespan. Verma and colleagues introduced the multi-sink energy-efficient clustering (MEEC) protocol to address this issue by employing multiple sinks around the network's periphery, facilitating direct communication links and reducing energy depletion [22]. While effective in mitigating energy holes, the necessity of multiple sinks raises both the operational costs and complexity of synchronization among these sinks. Exploring different methodologies, some researchers have advocated for a corona-based model, aiming to evenly distribute energy consumption across the network. However, the introduction of numerous corona rings significantly increases the network's computational demands. Notably, most existing studies have focused on scenarios where the sink is located within the network, an arrangement not always applicable in remote or unattended environments.

To offer a solution in both homogeneous and heterogeneous network settings, Liu and team suggested adjusting nodes' transmission ranges to balance energy usage across the network [23]. This approach, although innovative, did not incorporate clustering strategies, leading to an increased number of necessary transmissions. In an attempt to refine node selection processes, Sharmin et al. divided the network into wedges, merging them based on nodes' residual energy; however, the approach stumbled in efficiently selecting head nodes [24]. Zhao et al. adopted a clustering-based corona model with cluster head selection primarily based on nodal energy, which proved limited due to the oversight of other crucial parameters [25]. Mishra and colleagues introduced a probability distribution function for node distribution focusing on minimizing energy and coverage gaps, but the reliance on multihop communication without clustering led to a high transmission rate, making it inefficient [26].

Prabha and Selvan proposed the formation of uneven clusters to conserve energy, creating smaller clusters near the sink and larger ones further away to combat energy depletion. Despite its potential for energy conservation, this method did not fully address the imbalance in load distribution [27].

Verma et al. revisited the multi-sink strategy to simplify communications in unattended applications, yet acknowledged the consequent increase in network costs [28]. Naeimi et al. introduced a novel routing protocol for clustered sensor networks aimed at achieving balanced energy consumption. By dynamically redistributing communication loads, this protocol seeks to extend the network's operational life by preventing the premature exhaustion of nodes near the sink, demonstrating a significant reduction in energy hole formation through simulations [29]. Further, a dynamic grouping strategy was proposed to alleviate the energy hole dilemma, featuring the introduction of two gateway nodes and a novel approach to cluster head energy management, ensuring a more balanced energy distribution across the network. These advancements illustrate the continuous effort and diverse approaches within the research community to solve the energy hole problem in WSNs, highlighting the complexity of achieving efficient, long-lasting wireless sensor networks essential for a wide range of applications.

5.6   DATA DISSEMINATION IN CLUSTERED WSNS

WSNs consist of spatially distributed, autonomous sensors designed to monitor various physical or environmental conditions, making the dissemination of collected data a fundamental aspect of their functionality. Within the framework of clustered WSNs, sensor nodes are systematically organized into clusters. Each cluster is typically led by one or more cluster heads tasked with the aggregation and forwarding of data to a central sink node or base station, streamlining the process of data communication across the network [14].

The establishment of clusters is a pivotal initial step, where nodes self-organize based on parameters such as proximity, energy levels, and communication capacity to enhance network resource management. The selection of cluster heads is a critical decision influenced by factors including energy availability and node capabilities, underscoring their role in coordinating data collection and transmission within clusters. The practice of data aggregation within these clusters, whereby sensor nodes compile and consolidate information before passing it to the cluster head, plays a vital role in minimizing the volume of data transmitted. This, in turn, contributes to the conservation of energy and bandwidth resources [13].

Communication within clusters allows for the efficient exchange of aggregated data, with cluster heads possibly undertaking further processing to optimize energy use. Meanwhile, communication between clusters involves the transfer of aggregated data to the sink node, facilitating the relay of collected information to the monitoring station. Nonetheless, this inter-cluster communication is often limited by the nodes' energy and range capabilities, necessitating the implementation of effective routing protocols to ensure efficient data flow [23].

To navigate these challenges, various routing protocols have been developed for use in clustered WSNs, aimed at optimizing data transmission both within and across clusters. Notable examples include the low-energy adaptive clustering hierarchy (LEACH) [15], the hybrid energy-efficient distributed (HEED) clustering, and the threshold-sensitive energy-efficient sensor network (TEEN) protocol. These protocols are specifically designed to tackle the inherent challenges of WSNs, such as ensuring energy efficiency, scalability, and robustness, thereby facilitating timely and reliable data dissemination throughout the network.

5.6.1 Classification of Data Dissemination Techniques

Data dissemination in wireless sensor networks involves the efficient transmission of information from source nodes to a sink or base station. Various techniques can be classified based on their approaches, objectives, and considerations. Here are some key classifications:

Data-centric vs. query-centric: In the realm of WSNs, different strategies and techniques are employed for efficient data dissemination and communication. Two fundamental approaches include data-centric and query-centric methodologies. Data-centric approaches prioritize the dissemination of sensed data, ensuring that collected information reaches its destination effectively. Conversely, query-centric approaches involve responding to specific inquiries initiated by the sink node or users, tailoring data transmission according to requested information.

Direct vs. indirect communication: The other critical distinction lies between direct and indirect communication methods. In direct communication, nodes transmit data directly to the sink or intended recipient, minimizing transmission delays and energy consumption. On the other hand, indirect communication routes data through intermediate nodes before reaching the destination, which can enhance reliability but may introduce additional latency.

Multi-hop vs. single-hop: By considering the dynamic nature of WSNs, communication patterns can further be categorized as multi-hop or single-hop. Multi-hop communication involves transmitting data through multiple intermediate nodes before reaching the sink, facilitating long-distance communication and network scalability. In contrast, single-hop communication occurs when nodes transmit data directly to the sink, suitable for scenarios where nodes are within close proximity.

Energy-aware approaches: Energy efficiency is a paramount concern in WSNs, prompting the development of energy-aware approaches. These techniques take into account the energy levels of nodes to prolong the network's lifetime. Strategies such as energy-efficient routing, dynamic clustering, and selective data transmission help mitigate energy consumption and extend the operational lifespan of the network.

Location-based techniques: Moreover, location-based techniques leverage the spatial information of nodes for efficient data dissemination. Geographic routing protocols exemplify this category, utilizing node locations to make forwarding decisions and optimize data transmission paths based on physical proximity. These techniques harness spatial awareness to enhance communication efficiency and network performance in WSNs.

Thus, the data dissemination in clustered WSNs involves a hierarchical structure where sensor nodes aggregate and transmit data to cluster heads, which then relay the data within and between clusters. The choice of specific techniques and protocols depends on the application requirements, energy constraints, and the network architecture.

5.7 COMPARATIVE PERFORMANCE ANALYSIS OF THE CLUSTERED WSNS

In WSNs, the process of data dissemination is crucial for transmitting information from the sensors to a central sink or base station efficiently. This process is tailored through various strategies and protocols, each designed with specific goals and constraints in mind. Below is an overview of some primary classifications of these dissemination techniques:

- **Data-centric vs. query-centric approaches**: WSNs adopt either data-centric or query-centric approaches for data handling. Data-centric methods focus on the dissemination of all sensed data, aiming to ensure the comprehensive delivery of collected information to its intended destination. In contrast, query-centric strategies respond to specific requests from the sink node or users, with data transmission tailored to the needs of these inquiries.
- **Direct vs. indirect communication**: This classification differentiates between direct and indirect data transmission modes. Direct communication involves sensor nodes sending data straight to the sink or target recipient, which can reduce transmission delays and conserve energy. Indirect communication, however, involves routing data through one or more intermediate nodes, potentially increasing reliability at the cost of added latency.
- **Multi-hop vs. single-hop communication**: The distinction between multi-hop and single-hop communication addresses the network's capacity to handle data transmission over various distances. Multi-hop communication allows data to pass through several intermediate nodes, enhancing the network's reach and scalability. Single-hop communication, suitable for shorter distances, involves direct data transmission to the sink without intermediary nodes.
- **Energy-aware techniques**: Given the critical importance of energy conservation in WSNs, energy-aware techniques prioritize the efficient use of power. These methods include energy-efficient routing, dynamic clustering, and selective data transmission strategies designed to reduce energy consumption and extend network longevity.
- **Location-based strategies**: Utilizing the geographical positions of nodes, location-based strategies employ geographic routing protocols and other spatially aware techniques to optimize data transmission paths. By leveraging the physical locations of nodes, these techniques aim to improve the efficiency and performance of data dissemination in WSNs.

Data dissemination typically follows a hierarchical structure. Sensor nodes collect and aggregate data, then forward it to cluster heads, which subsequently manage the distribution of information both within and between clusters. The selection of dissemination techniques and protocols is guided by various factors, including the specific requirements of the application, energy constraints, and the overarching network architecture, ensuring effective and efficient communication throughout the network.

5.7.1　Network Lifetime Enhancement and Load Balancing

The advent of WSNs has marked a significant milestone in the technological revolution, largely due to their ability to monitor and collect data across various environments. Despite their versatility, one of the principal challenges facing WSNs is energy conservation, as the operational efficacy of sensor nodes is heavily dependent on battery life. These nodes, often deployed in inaccessible areas, rely on limited-capacity batteries, posing significant hurdles to both the longevity and efficiency of WSNs. Consequently, energy management becomes a pivotal concern, aiming to maximize network lifespan and performance.

In WSNs, the concept of network lifetime is closely linked to the operational status of alive nodes. This emphasizes the importance of maintaining a sufficient number of functioning nodes to ensure continuous network operation and data transmission capabilities. Figure 5.9 illustrates a comparative investigation of the number

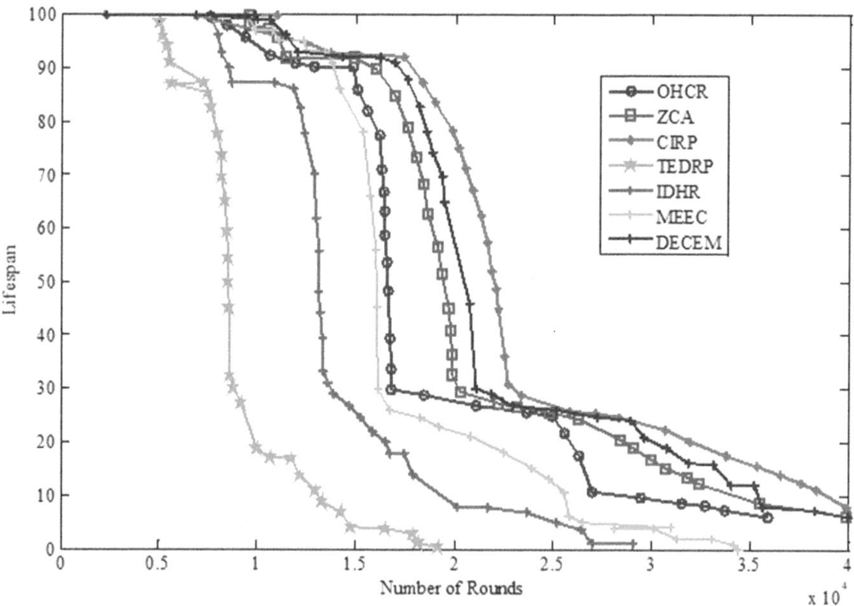

FIGURE 5.9　Number of alive nodes per round in various protocols

of alive nodes per round for the existing protocols DECEM [30], CIRP [31], ZCA [32], OHCR [33], MMEC [34], IDHR [35], and TEDRP [36]. The protocol DECEM enhances the network lifetime by 109.86%, 37.41%, 25.70%, 22.86%, 37.41%, 16.70%, and 10.23% in comparison to the TEDRP, IDHR, MEEC, OHCR, ZCA, and CIRP, respectively. The DECEM protocol revolves around single-hop communication over shorter distances. However, the limited scope in other protocols is a result of adopting a single sink approach, which affects the functionality of networks.

In addressing these challenges, researchers have proposed numerous strategies for energy conservation, focusing on optimizing energy use within sensor nodes (SNs) to extend the network's operational duration. Among these, software-defined networking (SDN) and clustering emerge as leading approaches for enhancing energy efficiency [37]. SDN, a modern network framework, separates the network's control logic from its physical components, such as switches or SNs. This separation allows network devices to focus solely on data forwarding, while the SDN controller oversees a broad spectrum of computational and managerial tasks [38]. The SDN architecture comprises three layers: the application, control, and data planes, with the latter consisting of passive devices managed by the control plane, facilitating efficient data packet management [39].

Incorporating SDN within WSNs, particularly through a multi-objective optimization model, has shown promise in elevating both energy efficiency and overall network performance. This model considers various factors, including packet delivery ratio, energy consumption, latency, and throughput, allowing for a tailored approach to routing protocol selection based on current network needs. However, this method faces scalability challenges due to the centralized controller's extensive responsibilities and the complexity of managing an extensive dataset of routing protocols [40].

Further innovations in energy-efficient routing include the use of fuzzy logic and game theory within the SDN framework [41]. Fuzzy logic approaches segment the network into zones, each managed by a separate controller, optimizing path selection by evaluating multiple parameters. Game theory models, on the other hand, focus on maximizing the residual energy of nodes and routing efficiency, promoting collaborative strategies among nodes. Despite their benefits, these methods can increase energy consumption due to the overhead associated with controller-node communication [42]. Traditional routing protocols often rely on isolated (distributed) strategies, focusing solely on the sender and receiver nodes without considering the energy status of neighbouring nodes. This approach can lead to suboptimal energy consumption patterns across the network. By contrast, global energy route saving optimization methods maintain comprehensive information on each node's status, including remaining energy and location, to inform routing decisions, aiming to minimize overall energy usage [43].

5.7.2 RESIDUAL ENERGY CONSUMPTION

In WSNs, the concept of residual energy consumption is pivotal, highlighting the energy that remains in sensor nodes after a period of use. These nodes, typically battery-operated, face limitations in energy supply, making efficient use a key determinant of the network's operational lifespan and functionality. The assessment of residual energy consumption serves as a crucial gauge for the network's energy efficiency and durability. This is quantified by deducting the energy expended over a given time from the sensor nodes' initial energy reserve. This evaluation aids operators and researchers in gauging the energy utilization efficiency of sensor nodes and the extent of energy available for ongoing operations. The oversight and optimization of this residual energy are fundamental for enhancing the network's energy efficiency, requiring the adoption of energy-conserving protocols, algorithms, and strategies to extend the network's life while preventing rapid energy depletion of the nodes.

Especially in applications like environmental monitoring or surveillance, where sensor nodes are placed in remote or hard-to-reach areas, effective energy management becomes indispensable. The average residual energy of a WSN post each iteration is calculated through the formula $RE_{Avg} = 1 / n\left(\sum_{i=1}^{N} RE_i\right)$, where N represents the total number of sensor nodes within the network and RE_i denotes the residual energy of the ith sensor node following the iteration. This formula averages the residual energy across all network nodes, offering a comprehensive overview of the energy still available in the network after certain operations. The precise computation of RE_i hinges on the WSN's energy model, incorporating initial energy, energy used during the operation, and any replenishment mechanisms, if present. The specific calculation details of RE_i will differ based on the employed energy model and the network's characteristics.

Moreover, the performance of the protocols is analyzed with a specific focus on the intricate dynamics of the average residual energy of the network. This analysis aims to unravel the nuances of how residual energy, or the remaining energy within sensor nodes, influences the overall efficiency and sustainability of WSNs. Moreover, the metric of remaining network energy is utilized to gauge the rate of energy consumption per round, as illustrated in Figure 5.10 for the protocols DECEM [30], CIRP [31], ZCA [32], OHCR [33], MMEC [34], IDHR [35], and TEDRP [36]. The protocol DECEM utilizes the nodes' energy efficiently in comparison to CIRP, ZCA, OHCR, MMEC, IDHR, and TEDRP. This enhanced performance of DECEM is the result of enabling cluster heads to consume less energy for packet transmission, thereby conserving the overall energy reservoir of nodes. Additionally, intra-cluster communication plays a pivotal role in preserving node energy by reducing the energy-hole issue reduction approach.

5.7.3 Optimal Cluster Head Selection Techniques

WSNs are composed of numerous small sensor nodes, each tasked with collecting and forwarding data to a central hub or base station. Positioned within a predefined area, these nodes self-organize to form a network capable of monitoring hard-to-reach environments. Despite their potential, the performance of these networks is often hampered by the limited energy, processing, and communication capabilities

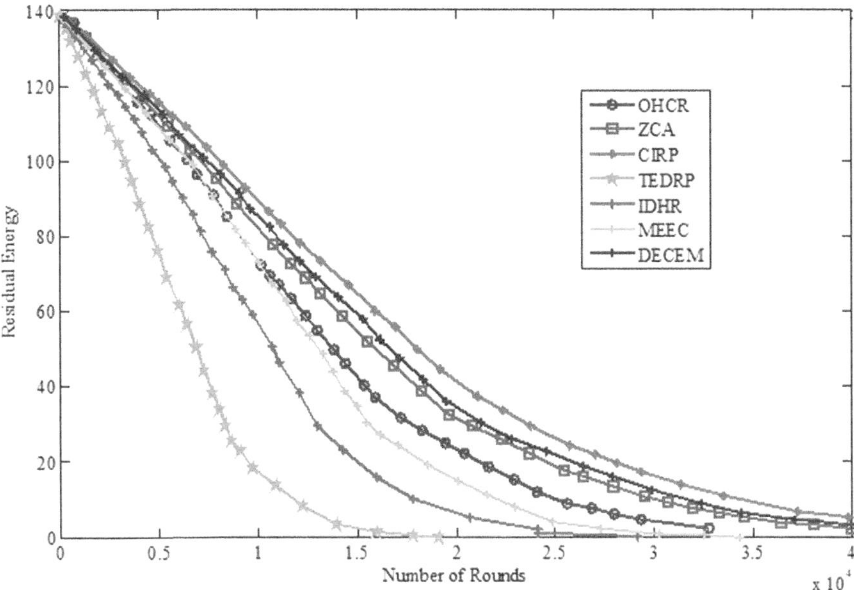

FIGURE 5.10   Energy consumption per round in various protocols

of the sensor nodes [30]. Given the challenge of replenishing the power sources of these nodes, it's crucial to minimize their energy consumption to extend the network's overall lifespan [31]. This necessity has sparked a wave of research aiming to enhance network efficiency and longevity [32]. Addressing this challenge, the technique of selecting cluster heads has emerged as a pivotal strategy for network optimization [32]. Clustering, as a technique, stands out for its ability to enhance network performance, reduce energy demands, and increase the network's operational duration [34]. The process of choosing cluster heads is critical because these nodes manage cluster communications and forward aggregated data to the sink node, significantly influencing the network's energy efficiency and longevity.

Cluster head selection is vital for several reasons. First, it directly affects energy conservation within the network, as cluster heads handle the dual tasks of data aggregation and communication with the sink node. Selecting energy-efficient nodes for these roles helps preserve the network's energy resources, extending its operational life [35]. Moreover, by aggregating data locally, cluster heads reduce the need for long-distance data transmission, which in turn conserves energy and eases network congestion. Additionally, the strategic selection of cluster heads is key to ensuring the network's longevity. Well-balanced energy usage among nodes prevents early energy depletion, thereby sustaining the network's functionality over a longer period. Cluster heads also streamline network operations by managing intra-cluster communications and reducing control and data packet overhead. This efficiency not only optimizes data routing but also minimizes redundant communications, further enhancing network performance.

Recent studies have introduced innovative approaches to cluster head selection. Abbas et al. [44] proposed the HQCA method, which aims to produce high-quality clusters by evaluating cluster quality and optimizing distances within and between clusters using fuzzy logic. This approach considers several factors, including node energy levels and distances, to select the optimal cluster head. Rehman et al. [45] introduced a secure cluster head selection method, which calculates node weight by incorporating a trust metric, ensuring both security and energy efficiency. Al-Kaseem et al. [46] developed the SEA model to reduce message exchanges between nodes and avoid frequent changes in cluster head positions, while Sindhanaiselvan et al. [16] focused on minimizing energy consumption through an efficient cluster head selection process, directly impacting the network's lifespan. Thus, the choice of an optimal cluster head selection strategy is a cornerstone of efficient WSN operation. Tailoring the selection process to meet the unique demands and constraints of specific WSN applications can markedly improve energy efficiency, data aggregation, and network longevity. Ongoing research and development efforts are key to refining these strategies, ensuring that WSNs can meet the challenges of diverse application scenarios and maximize both performance and operational lifespan.

5.7.4 Throughput Optimization in Clustered WSNs

Optimizing throughput in clustered WSNs is essential for improving data transfer rates and ensuring the efficient use of resources. In such networks, nodes are grouped into clusters, each managed by a cluster head chosen based on criteria such as remaining energy, distance to the sink node, and communication capabilities. The strategic selection and positioning of cluster heads are crucial as they coordinate the data transmission from the cluster to the sink, significantly influencing the network's throughput. Routing protocols are another vital component in enhancing throughput. These protocols dictate the path data takes from its origin to the sink node. Opting for protocols that focus on conserving energy, evenly distributing network load, and preventing data congestion can lead to notable increases in throughput. Furthermore, aggregating data at the cluster head level before it's sent to the sink node helps reduce the volume of data transmission, thereby improving throughput. This process involves merging similar or repetitive data from within the cluster, which decreases the total number of packets sent and conserves network bandwidth.

To further enhance throughput, scheduling mechanisms and time division multiple access (TDMA) can be employed to prevent data collisions and ensure efficient channel usage. Assigning specific time slots for data transmission to nodes or clusters reduces interference and boosts throughput. Cross-layer optimization represents a comprehensive strategy that involves making coordinated decisions across different layers of the protocol stack. For example, adjusting routing decisions based on live data from the physical layer can lead to significant throughput improvements. Load balancing, which distributes the data load evenly across the network, helps avoid congestion in certain areas, promoting more efficient resource use and maintaining high throughput levels.

Addressing the multifaceted challenge of throughput optimization in clustered WSNs requires a balanced approach that considers energy efficiency, network longevity, and reliability. The development and refinement of algorithms and protocols continue to play a pivotal role in enhancing clustered WSN performance across various applications. Notably, improving throughput also contributes to the network's energy efficiency, which is directly linked to its operational lifespan. Efficient data transmission methods that minimize energy consumption are crucial, especially in scenarios where sensor nodes are deployed in remote or hard-to-access locations, making battery replacement impractical.

Additionally, throughput is a fundamental metric that helps to ensure that the protocols meet the required data transfer rate in the network. Figure 5.11 illustrates the quantity of data packets transmitted to the sink per round in homogeneous networks for the existing protocols DECEM [30], CIRP [31], ZCA [32], OHCR [33], MMEC [34], IDHR [35], and TEDRP [36]. The protocol DECEM enhances the throughput by 149.82%, 67.31%, 55.70%, 42.86%, 25.41%, 17.24%, and 8.23% in comparison to the TEDRP, IDHR, MMEC, OHCR, ZCA, and CIRP, respectively. This enhancement is attributed to optimized cluster head selection considering energy consumption and distance optimization, thereby enhancing network durability.

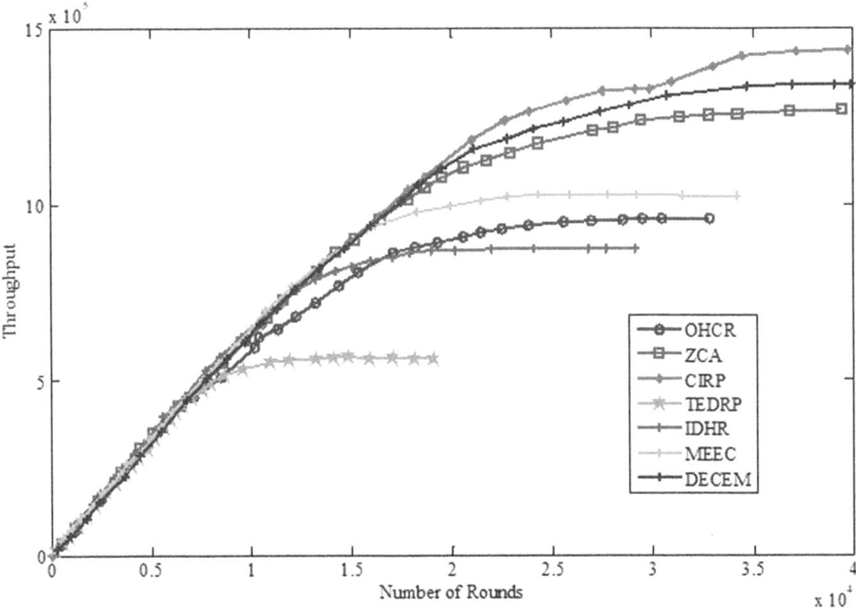

FIGURE 5.11 Data packets transmitted to sink per round in different protocols

5.8 CONCLUSION

The comprehensive exploration of energy-efficient clustering in WSNs underscores its vital role in optimizing network performance and extending lifespan. Through detailed discussions ranging from the foundational principles of clustering to advanced cluster head selection algorithms and mitigation strategies for the energy hole problem, this chapter has illuminated the multifaceted benefits of clustering in enhancing the operational efficiency of WSNs. Clustering emerges as a strategic approach to managing the inherent challenges of WSNs, particularly the limited energy resources of sensor nodes. By organizing nodes into clusters and designating CHs, networks achieve significant improvements in energy efficiency, data dissemination, and overall communication effectiveness. This organizational structure not only facilitates the reduction of energy consumption but also supports the equitable distribution of workload among sensor nodes, thereby prolonging network lifetime.

The exploration of both metaheuristic-based and traditional CH selection algorithms reveals the dynamic landscape of strategies available to network designers. Metaheuristic algorithms, inspired by natural processes, offer robust solutions for identifying optimal CHs, showcasing the potential for adaptive and intelligent network management. Conversely, non-metaheuristic methods provide simpler, yet effective, approaches for CH selection, emphasizing the versatility of clustering strategies to accommodate diverse network requirements and constraints. The chapter's in-depth analysis of the energy hole problem further highlights the critical importance of strategic clustering in maintaining network integrity. Through the identification of key characteristics and the introduction of targeted mitigation techniques, it becomes evident that effective clustering can significantly alleviate the challenges posed by uneven energy consumption across the network.

Moreover, the comparative performance analysis of clustered WSNs, focusing on network lifetime enhancement, load balancing, residual energy consumption, optimal CH selection techniques, and throughput optimization, provides valuable insights into the practical implications of clustering. This analysis demonstrates that energy-efficient clustering not only contributes to the sustainability of WSNs but also enhances their capability to support diverse applications, from environmental monitoring to surveillance. Thus, energy-efficient clustering stands as a cornerstone of WSN design and operation, offering a comprehensive framework for addressing the critical challenges of energy consumption and network performance. The insights and strategies presented in this chapter underscore the necessity of adopting a holistic approach to clustering, one that integrates advanced selection algorithms, energy hole mitigation techniques, and efficient data dissemination practices. As WSNs continue to evolve and support increasingly complex applications, the principles and methodologies of energy-efficient clustering will remain pivotal in unlocking their full potential, ensuring their sustainability, and maximizing their impact across various domains.

REFERENCES

1. Al-Sulaifanie, A. I., B. K. Al-Sulaifanie, and S. Biswas. "Recent Trends in Clustering Algorithms for Wireless Sensor Networks: A Comprehensive Review." *Computer Communications* 191 (2022): 395–424.
2. Singh, S. et al. "A Genetic Algorithm Based Dynamic Transmission of Data for Communicable Disease in IoMT Environment." *IEEE Internet of Things Journal* (2023). https://doi.org/10.1109/jiot.2023.3288614.
3. Singh, S. et al. "An Energy-Efficient Modified Metaheuristic Inspired Algorithm for Disaster Management System Using WSNs." *IEEE Sensors Journal* 21, no. 13 (2021): 15398–408.
4. Nandan, A. S. et al. "An Optimized Genetic Algorithm for Cluster Head Election Based on Movable Sinks and Adjustable Sensing Ranges in IoT-based HWSNs." *IEEE Internet of Things Journal* 9, no. 7 (2021): 5027–39.
5. Nandan, A. S. et al. "A Green Data Collection & Transmission Method for IoT-based WSN in Disaster Management." *IEEE Sensors Journal* 21, no. 22 (2021): 25912–21.
6. Singh, S. et al. "A GA-Based Sustainable and Secure Green Data Communication Method Using IoT-Enabled WSN in Healthcare." *IEEE Internet of Things Journal* 9, no. 10 (2021): 7481–90.
7. Singh, S. "A Threshold-Based Optimization Energy-Efficient Routing Technique in Heterogeneous Wireless Sensor Networks." *Machine Learning and Cognitive Computing for Mobile Communications and Wireless Networks* (2020): 203–24. https://doi.org/10.1002/9781119640554.ch9.
8. Singh, S. et al. "A Secure Energy-Efficient Routing Protocol for Disease Data Transmission Using IoMT." *Computers and Electrical Engineering* 101 (2022): 108113.
9. Singh, S. et al. "A Proficient Data Gathering Technique for Unmanned Aerial Vehicle-Enabled Heterogeneous Wireless Sensor Networks." *International Journal of Communication Systems* 34, no. 16 (2021): e4956.
10. Nandan, A. S. et al. "A Green Data Collection & Transmission Method for IoT-Based WSN in Disaster Management." *IEEE Sensors Journal* 21, no. 22 (2021): 25912–21.
11. Singh, S., D. Garg, and A. Malik. "A Novel Cluster Head Selection Algorithm based IoT Enabled Heterogeneous WSNs Distributed Architecture for Smart City." *Microprocessors and Microsystems* (2023): 104892. https://doi.org/10.1016/j.micpro.2023.104892.
12. Benelhouri, A., H. Idrissi-Saba, and J. Antari. "An Improved Gateway-Based Energy-Aware Multi-Hop Routing Protocol for Enhancing Lifetime and Throughput in Heterogeneous WSNs." *Simulation Modelling Practice and Theory* 116 (2022): 102471.
13. Raslan, A. F. et al. "An Improved Sunflower Optimization Algorithm for Cluster Head Selection in the Internet of Things." *IEEE Access* 9 (2021): 156171–86.
14. Cheng, H. et al. "Nodes Organization for Channel Assignment with Topology Preservation in Multi-Radio Wireless Mesh Networks." *Ad Hoc Networks* 10, no. 5 (2012): 760–73.
15. Heinzelman, W. R., A. Chandrakasan, and H. Balakrishnan. "Energy-Efficient Communication Protocol for Wireless Microsensor Networks." In Proceedings of the 33rd Annual Hawaii International Conference on System Sciences, p. 10-pp, 2000.
16. Sindhanaiselvan, K., J. Mannar Mannan, and S. K. Aruna. "Designing a Dynamic Topology (DHT) for Cluster Head Selection in Mobile Adhoc Network." *Mobile Networks and Applications* 25 (2020): 576–84.
17. Zhou, W. "Energy Efficient Clustering Algorithm Based on Neighbors for Wireless Sensor Networks." *Journal of Shanghai University* 15, no. 2 (2011): 150–3.

18. Pachlor, R., and D. Shrimankar. "LAR-CH: A Cluster-Head Rotation Approach for Sensor Networks." *IEEE Sensors Journal* 18, no. 23 (2018): 9821–8.
19. Tyagi, V., and S. Singh. "Network Resource Management Mechanisms in SDN Enabled WSNs: A Comprehensive Review." *Computer Science Review* 49 (2023): 100569.
20. Edla, D. R., M. C. Kongara, and R. Cheruku. "SCE-PSO Based Clustering Approach for Load Balancing of Gateways in Wireless Sensor Networks." *Wireless Networks* 25, no. 3 (2019): 1067–81.
21. Edla, D. R., M. C. Kongara, and R. Cheruku. "A PSO Based Routing with Novel Fitness Function for Improving Lifetime of WSNs." *Wireless Personal Communications* 104, no. 1 (2019): 73–89.
22. Verma, S., N. Sood, and A. K. Sharma. "A Novelistic Approach for Energy Efficient Routing Using Single and Multiple Data Sinks in Heterogeneous Wireless Sensor Network." *Peer-to-Peer Networking and Applications* 12 (2019): 1110–36.
23. Liu, X. "A Novel Transmission Range Adjustment Strategy for Energy Hole Avoiding in Wireless Sensor Networks." *Journal of Network and Computer Applications* 67 (2016): 43–52.
24. Sharmin, N., A. Karmaker, W. L. Lambert, M. S. Alam, and M. S. T. Shawkat. "Minimizing the Energy Hole Problem in Wireless Sensor Networks: A Wedge Merging Approach." *Sensors* 20, no. 1 (2020): 277.
25. Zhao, X., X. Xiong, Z. Sun, X. Zhang, and Z. Sun. "An Immune Clone Selection Based Power Control Strategy for Alleviating Energy Hole Problems in Wireless Sensor Networks." *Journal of Ambient Intelligence and Humanized Computing* 11 (2020): 2505–18.
26. Mishra, R., V. Jha, R. K. Tripathi, and A. K. Sharma. "Corona Based Node Distribution Scheme Targeting Energy Balancing in Wireless Sensor Networks for the Sensors Having Limited Sensing Range." *Wireless Networks* 26 (2020): 879–96.
27. Prabha, K. L., and S. Selvan. "Energy Efficient Energy Hole Repelling (EEEHR) Algorithm for Delay Tolerant Wireless Sensor Network." *Wireless Personal Communications* 101, no. 3 (2018): 1395–409.
28. Verma, S., N. Sood, and A. K. Sharma. "Genetic Algorithm-Based Optimized Cluster Head Selection for Single and Multiple Data Sinks in Heterogeneous Wireless Sensor Network." *Applied Soft Computing* 85 (2019): 105788.
29. Nandan, A. S., S. Singh, and L. K. Awasthi. "An Efficient Cluster Head Election Based on Optimized Genetic Algorithm for Movable Sinks in IoT Enabled HWSNs." *Applied Soft Computing* 107 (2021): 107318.
30. Dogra, R., S. Rani, B. Sharma, S. Verma, D. Anand, and P. Chatterjee. "A Novel Dynamic Clustering Approach for Energy Hole Mitigation in Internet of Things-Based Wireless Sensor Network." *International Journal of Communication Systems* 34, no. 9 (2021): e4806.
31. Pokhrel, S. R., S. Verma, S. Garg, A. K. Sharma, and J. Choi. "An Efficient Clustering Framework for Massive Sensor Networking in Industrial Internet of Things." *IEEE Transactions on Industrial Informatics* 17, no. 7 (2020): 4917–24.
32. Mohemed, R. E., A. I. Saleh, M. Abdelrazzak, and A. S. Samra. "Energy-Efficient Routing Protocols for Solving Energy Hole Problem in Wireless Sensor Networks." *Computer Networks* 114 (2017): 51–66.
33. Benelhouri, A., H. Idrissi-Saba, and J. Antari. "An Improved Gateway-Based Energy-Aware Multi-Hop Routing Protocol for Enhancing Lifetime and Throughput in Heterogeneous WSNs." *Simulation Modelling Practice and Theory* 116 (2022): 102471.

34. Banerjee, J., S. K. Mitra, P. Ghosh, and M. K. Naskar. "Memory Based Message Efficient Clustering (MMEC) for Enhancement of Lifetime in Wireless Sensor Networks Using a Node Deployment Protocol." *In Proceedings of 2011 International Conference Communication, Computing & Security* (2011): 71–76. https://doi.org/10.1145/1947940.1947956.

35. Verma, S., N. Sood, and A. K. Sharma. "A Novelistic Approach for Energy Efficient Routing Using Single and Multiple Data Sinks in Heterogeneous Wireless Sensor Network." *Peer-to-Peer Networking and Applications* 12 (2019): 1110–36.

36. Mittal, N., U. Singh, and B. S. Sohi. "A Novel Energy Efficient Stable Clustering Approach for Wireless Sensor Networks." *Wireless Personal Communications* 95 (2017): 2947–71.

37. Rojas, E. et al. "Are We Ready to Drive Software-Defined Networks? A Comprehensive Survey on Management Tools and Techniques." *ACM Computing Surveys* 51, no. 2 (2018): 1–35.

38. Zhu, L. et al. "SDN Controllers: A Comprehensive Analysis and Performance Evaluation Study." *ACM Computing Surveys* 53, no. 6 (2020): 1–40.

39. Tyagi, V., and S. Singh. "GM-WOA: A Hybrid Energy Efficient Cluster Routing Technique for SDN-Enabled WSNs." *The Journal of Supercomputing* 2023 (2023): 1–29.

40. Misra, S., S. Bera, A. M. P., S. K. Pal, and M. S. Obaidat. "Situation-Aware Protocol Switching in Software-Defined Wireless Sensor Network Systems." *IEEE Systems Journal* 12, no. 3 (2018): 2353–60.

41. Banerjee, A., and D. Hussain. "SD-EAR: Energy Aware Routing in Software Defined Wireless Sensor Networks." *Applied Sciences* 8, no. 7 (2018): 1013.

42. Hassan, A., A. Anter, and M. Kayed. "A Survey on Extending the Lifetime for Wireless Sensor Networks in Real-Time Applications." *International Journal of Wireless Information Networks* 28, no. 1 (2021): 77–103.

43. Huang, H., Z. Wu, and S. Tang. "Energy-Saving Route Optimization in a Software-Defined Wireless Sensor Network." *International Journal of Distributed Sensor Networks* 14, no. 10 (2018): 1550147718807655.

44. Sengathir, J., A. Rajesh, G. Dhiman, S. Vimal, C. A. Yogaraja, and W. Viriyasitavat. "A Novel Cluster Head Selection Using Hybrid Artificial Bee Colony and Firefly Algorithm for Network Lifetime and Stability in WSNs." *Connection Science* 34, no. 1 (2022): 387–408.

45. Rehman, E., M. Sher, S. H. A. Naqvi, K. B. Khan, and K. Ullah. "Energy Efficient Secure Trust Based Clustering Algorithm for Mobile Wireless Sensor Network." *Journal of Computer Networks and Communications* 2017 (2017). https://doi.org/10.1155/2017/1630673.

46. Al-Kaseem, B. R., Z. K. Taha, S. W. Abdulmajeed, and H. S. Al-Raweshidy. "Optimized Energy–Efficient Path Planning Strategy in WSN with Multiple Mobile Sinks." *IEEE Access* 9 (2021): 82833–47.

Chapter 6

SUMMARY

Chapter 6 discusses the diverse approaches and methodologies within wireless sensor networks (WSNs) for routing, focusing on those with adjustable sensing ranges. The chapter begins by introducing the fundamental routing methods in WSNs and then classifies these protocols based on network structure, orientation, and the initiators of communication. The chapter highlights both the advantages, such as improved efficiency and network longevity, and the limitations, including scalability issues and energy constraints, inherent in WSN routing protocols. A significant portion of the chapter is dedicated to Variable Sensing Range (VaseRa) and its influence on WSN performance, illustrating how adjustable sensing ranges can enhance network adaptability and efficiency. The text delves into energy-efficient metaheuristics-based routing protocols, specifically optimized genetic algorithms (GAs) for cluster head (CH) selection. This section details GA workflow, including fitness evaluation for CH selection, cluster formation techniques, and data collection strategies involving sinks and sensor nodes. The performance analysis of these GA-based routing methods is thoroughly examined, comparing scenarios with static and multiple sinks, and assessing the impact of movable sinks with adjustable sensing ranges on network lifetime, energy consumption, and throughput. The conclusion synthesizes the findings, emphasizing the importance of adjustable sensing ranges and energy-efficient routing protocols in optimizing WSN performance and sustainability.

DOI: 10.1201/9781003427780-11

6 Routing Methods with Adjustable Sensing Range

6.1 INTRODUCTION TO ROUTING METHODS IN WSNS

In the ambit of wireless sensor networks, efficient and reliable data communication is of extreme importance between sensor nodes (SN). These networks consist of numerous sensor nodes with limited power supply and computational resources. Wireless sensor networks (WSNs) offer low-cost solutions to problems such as war surveillance, environment monitoring, precision agriculture, industrial automation, traffic management, healthcare, wildlife tracking, disaster management, water quality monitoring, structural health monitoring, object tracking, and habitat monitoring. The resource constraints and dynamic nature of these networks pose unique challenges in terms of efficient energy dissipation, network scalability, and adaptability, necessitating innovative routing solutions. This necessity has urged the requirement for innovative routing solutions to optimize data transmission and ensure the longevity and optimal performance of WSNs [1].

To address these challenges, an energy-efficient information collection process is essential to reduce energy consumption. One such method is clustering, where nodes are organized into groups. In clustering, after the deployment of the sensor nodes in the monitoring area, the cluster heads (CHs) are elected among the SNs, which collect information from the respective cluster members and forward it to the sink through other CHs or directly. The cluster nodes send data directly to the sink, which is called one-hop communication, whereas the CHs send data through multiple cluster nodes to the sink, which is called multi-hop communication. The election of the CH is an iterative process based on the network parameters. An effective routing method plays a pivotal role in reducing power consumption along with latency. These methods are categorized based on their design and functionalities, addressing the unique requirements and constraints of WSNs [2]. The routing protocols classification is discussed in the next section.

6.2 CLASSIFICATION OF ROUTING PROTOCOLS

Historically, a multitude of routing techniques has been put forward to facilitate efficient and dependable data transmission in WSNs [3]. These routing protocols for WSNs can be extensively categorized by examining a variety of criteria. These criteria include the structure of the network, the methodologies for establishing pathways, the foundational principles of the protocols, and the initiators of communication.

DOI: 10.1201/9781003427780-12

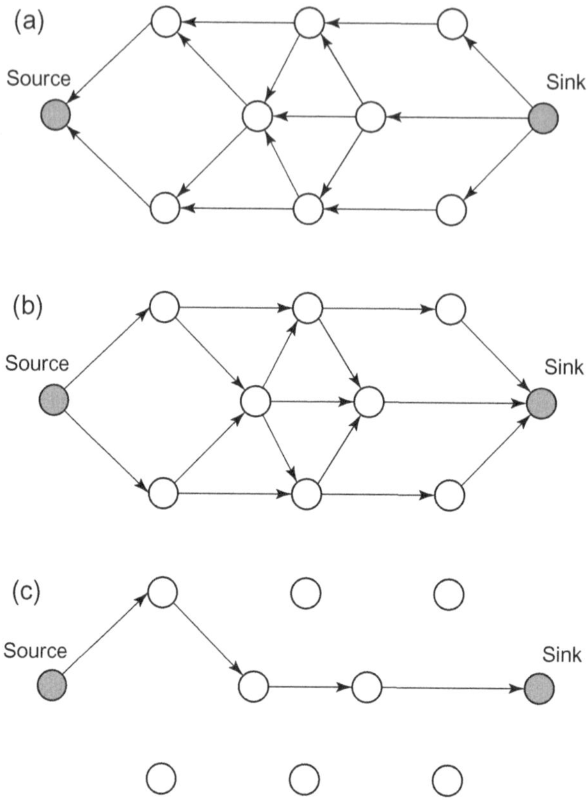

FIGURE 6.1 Routing protocol classification. (a) Propagate interest, (b) set up gradients, (c) send data and path reinforcement

This broad classification is visually delineated in Figure 6.1, providing a clear framework for understanding the different dimensions and approaches utilized in routing within wireless sensor networks.

6.2.1 Network Structure-Based Routing

In the context of network architecture, it's possible to categorize networks into three main types: flat, hierarchical, and location-based. Within a flat network setup, every node operates on an equal level, carrying out identical roles and possessing with the same capabilities. Given the extensive scale of nodes present, it becomes unfeasible to allocate a unique identifier to each node. This challenge has paved the way for the evolution of information-centric routing approaches. In such methodologies, the base station (BS) dispatches requests to designated areas and stands by for the retrieval of information from sensors located in those specified zones. Attribute-based naming becomes essential in this context to specify data properties. Early protocols like

SPIN and directed diffusion demonstrated energy savings through data exchange and redundant data elimination, inspiring the creation of similar protocols.

6.2.1.1 Flat Routing

A widely recognized approach that focuses on a data-centric, application-aware information consolidation technique for WSNs is known as directed diffusion and was proposed by C. Intanagonwiwat [4]. This method integrates information while it is being transmitted from diverse sources, aiming to minimize the total number of data transmissions and remove any duplicate data.

In the directed diffusion model, sensors report occurrences, creating gradients of data within their localities. The BS initiates data requests through interest broadcasts, with interests representing tasks for the network. Interests propagate through the network hop-by-hop as each node broadcasts them to neighbours. Gradients are initialized during this process to retrieve the data fulfilling the enquiry towards the BS. When sensors detect a request, they form a data path leading back to the sensor nodes that initiated the request. This process is repeated, building data pathways from the data origin points to the central hub or BS. Gradients specify attribute values and directions, with varying strengths towards different neighbours influencing information flow. The system does not initially check for loops, addressing them at a later stage. The process involves building gradients, sending interests, and disseminating information. The information flows from multiple paths, with the best paths strengthened to prevent further flooding based on a local rule. To decrease the cost of communications, information is consolidated en route, aiming to find an efficient aggregation tree guiding data from source nodes to the sink. Periodic interest refreshes by the BS are necessary, as interests may not reliably transmit throughout the network. Figure 6.2 illustrates the working of directed diffusion in three stages: transmitting queries, establishing gradients, and distributing information.

In networks employing directed diffusion, each sensor node is equipped with application-specific knowledge, which contributes to energy conservation by facilitating the choice of the most efficient routes, along with the storage and handling of data internally. This strategy leverages caching to enhance the effectiveness, resilience, and scalability of communication between sensor nodes, in harmony with the foundational ideals of the data diffusion model. Furthermore, directed diffusion is adept at automatically disseminating vital information to targeted areas within the sensor network. This method of data gathering is especially effective for ongoing inquiries, where the nodes making requests expect to receive data that meets their criteria over a prolonged period. Conversely, it is not considered appropriate for single-use queries because the initial effort required to establish directional paths for one-time use does not justify the anticipated advantages.

The success of data consolidation techniques within the directed diffusion framework is shaped by several variables, such as the locations of the originating nodes, the quantity of these nodes, and the structure of the communication network. To delve into these variables, two models for placing sources were examined, namely the random sources (RS) model and the event radius (ER) model, as shown in Figure 6.3. The ER model designates a specific point within the network space as an event

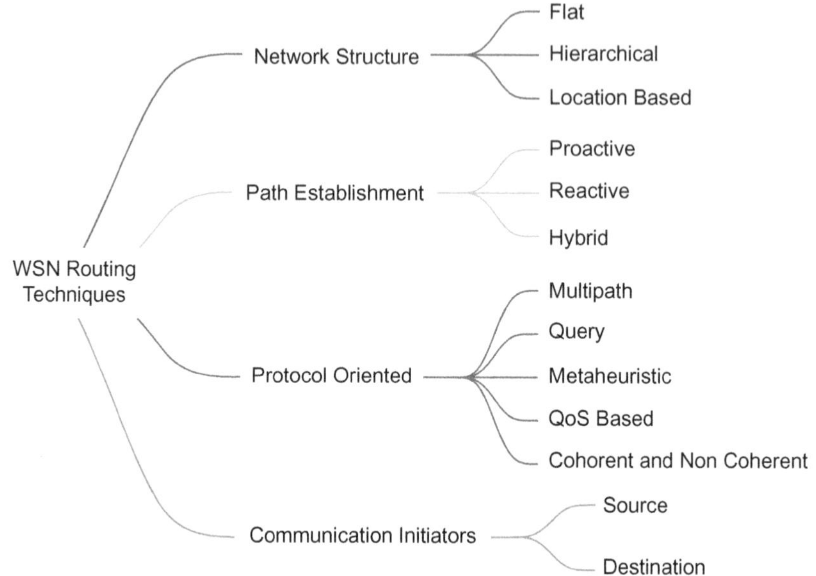

FIGURE 6.2 An example of interest diffusion in a sensor network: (a) event radius model, (b) random source model

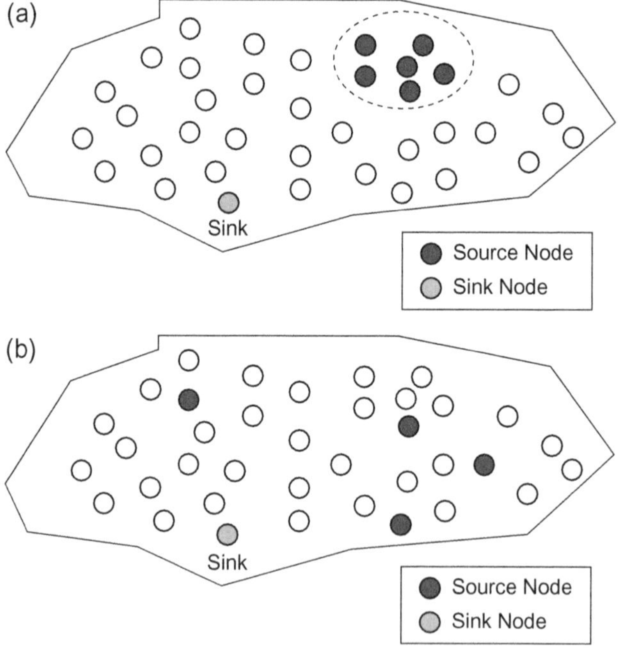

FIGURE 6.3 Models used in data-centric routing paradigm

location, and nodes within a certain sensing distance S are identified as data sources. The expected density of sources in a network zone with n sensor nodes is roughly $\pi S^2 n$. Conversely, the RS model selects k nodes at random, other than the BS nodes, to act as sources, without necessitating any grouping. Both approaches facilitate connecting a larger number of sources to the BS under the same energy constraints. Nevertheless, the impact on energy usage varies depending on the specific application in question. To sum up, the efficiencies gained through the aggregation process in directed diffusion can bolster network resilience against changes in the environmental conditions being monitored.

6.2.1.2 Hierarchical Routing

Hierarchical routing, initially developed for wired networks, has emerged as a highly effective strategy for enhancing scalability and communication efficiency, now adapted for use in WSNs to promote energy-efficient data transmission. In this structured framework, the nodes with higher energy reserves are tasked with information processing and forwarding duties, whereas those with less energy are assigned to monitor their immediate environment. The creation of clusters and the designation of specific responsibilities to cluster leaders significantly improve the network's scalability, operational lifespan, and energy utilization. By conducting data consolidation and fusion within clusters, hierarchical routing substantially cuts down on the energy expended in message transmission to the BS. This dual-layered routing technique encompasses one layer for the election of CHs and another for the routing process itself. Although it falls under the umbrella of hierarchical routing, many methodologies in this field extend beyond mere routing to include decisions on the timing and entities responsible for sending, processing, or aggregating data, along with channel allocation, among other aspects. These components, while distinct, complement the core multi-hop routing functionality.

In hierarchical routing, nodes with higher energy reserves are designated for processing and transmitting information to the central collection point, while those with less energy are assigned to the role of data collection. This approach indicates that the network is segmented into clusters, with a CH being chosen within each cluster. Such a strategy enhances network scalability, prolongs the network's operational life, and improves energy efficiency [1]. In the location-based routing technique, the originating node embeds the destination's location as the sink in every data packet. Subsequently, the intermediary nodes forward these packets to the neighbouring node that is nearest to the intended destination [2].

The low energy adaptive clustering hierarchy (LEACH), developed by Heinzelman et al. [5], is a protocol specifically crafted for sensor networks embodying a hierarchical clustering methodology. This protocol stands out for its cluster-based approach, where it autonomously organizes sensor nodes into distributed clusters. In a distinctive move, LEACH dynamically selects sensor nodes to serve as CHs, systematically rotating this responsibility to balance the energy consumption throughout the network. These CHs are tasked with compressing data collected from their cluster and forwarding this consolidated information to the BS, thereby streamlining the volume of data transmission.

LEACH incorporates a dual-mode medium access control (MAC) strategy, employing time division multiple access (TDMA) and code division multiple access (CDMA) techniques to significantly reduce the chances of data collision both within and between clusters. This protocol is particularly adept at centralized and periodic data gathering, making it ideal for continuous environmental monitoring. However, its design might not be as efficient for scenarios that do not require immediate or constant data forwarding. The protocol's innovative mechanism for rotating the CH's role promotes an equitable distribution of energy expenditure across the sensor network, enhancing its longevity and efficiency. Simulations have indicated that designating merely 5% of the nodes as CHs can lead to optimal energy management within the network.

LEACH protocol operates through two main stages: the setup and the steady state phases. During the setup phase, the network forms clusters and selects CHs, setting the stage for efficient data communication. The subsequent steady state phase, which occupies a more extended period, is when information is transmitted to the BS, thereby reducing the frequency of organizational overhead. In the initial setup phase, a specific proportion of nodes, referred to as "f", volunteer to be CHs in a stochastic manner. Each node generates a random number, "ran", within the range of 0 to 1. A node becomes a CH for the current cycle if its "ran" falls below a calculated threshold, $T(n)$. This threshold is calculated using a formula that accounts for the preferred ratio of CHs, the current cycle, and the set of nodes, labelled as G, that have not served as CHs in the previous $(1/f)$ cycles. This method promotes a balanced and adaptable selection of CHs, embodying the dynamic nature of the LEACH protocol's operation.

$$T(n) = \frac{f}{1 - f\left(ran\ mod\left(1/f\right)\right)} \quad if\ n\epsilon\ G \tag{6.1}$$

where G represents the collection of nodes eligible for participation in the selection process of CH. After being chosen, each CH sends out a notification broadcast to alert the remaining nodes in the network of their newly assigned leadership role. Nodes that are not CHs, upon receiving this notification, evaluate which cluster they prefer to join, basing their decision on the signal strength of the broadcasted message. These nodes then notify the respective CH of their desire to be part of their cluster.

Following the reception of all requests from nodes intending to join, a CH assesses the total membership of its cluster. With this information, the CH formulates a TDMA schedule, allocating specific time slots to each node for their data transmissions. This TDMA schedule is subsequently shared with all the nodes within the cluster. This process ensures that communications within the cluster are well-coordinated and occur without interference, allowing for efficient data sharing and reducing the risk of data collision.

During the steady state phase of the LEACH protocol, sensor nodes begin gathering and sending information to their designated CHs. Once the CHs receive this data, they consolidate it before sending it onward to the BS. After a set period has elapsed, the network reverts to the setup phase to start a new cycle of CH selection. It is important to highlight that each cluster employs unique CDMA codes for their communications. This strategy helps prevent interference from nodes belonging to different clusters. Through this repetitive cycle, the protocol achieves efficient and orderly data handling across the network while effectively reducing the potential for interference.

Virtual grid architecture routing introduces an energy-efficient routing strategy aimed at extending the lifespan of networks through strategic data consolidation and processing within the network [6]. This approach leverages the typically stationary or minimally mobile nature of nodes in various WSN applications, opting for a stable topology [7]. By adopting a method that does not rely on Global Positioning System (GPS) technology, the strategy establishes clusters characterized by uniformity, adjacency, and distinct boundaries, thus forming symmetrical configurations. Specifically, square-shaped clusters are utilized to construct a precise rectilinear virtual layout [6].

In each designated area, a node is selected based on optimal criteria to act as the CH, facilitating information aggregation at both the local and broader network scales. Local aggregators (LAs) are responsible for compiling data within their vicinity, and a selection of these LAs is further engaged for the network-wide aggregation tasks. Identifying the most effective master aggregators (MAs) for this wider scale aggregation presents a complex challenge, often described as an NP-hard problem. Illustrations, such as in Figure 6.4, can visually represent this structured zoning and the dual-layer aggregation mechanism inherent to the virtual grid architecture, showcasing the efficiency of this layout. Noteworthy is the flexibility in the positioning of the BS within this grid system. It does not have to be located at the periphery but can be situated anywhere within the network to optimize connectivity and data flow.

Routing strategies within networks are broadly categorized into three types: proactive, reactive, and hybrid. In a proactive routing approach, every node in the network keeps a table that lists paths to every other node, ensuring immediate route availability. Conversely, reactive routing strategies establish routes as needed. This means that when a node intends to send data to another, it initiates a process to discover a route to the destination node, thereby avoiding the constant upkeep of comprehensive routing tables and reducing network overhead. The hybrid routing approach combines elements of both proactive and reactive methods. Nodes in a hybrid system maintain routing information for certain destinations but resort to discovering routes on demand for communicating with the rest of the network. This blend aims to balance the immediate availability of route information with the efficiency of on-demand route discovery.

The challenge of integrating routing with data consolidation in networks is addressed through various strategic approaches. Initially, a solution incorporating an Integer Linear Program (ILP) offers a precise algorithmic framework, complemented

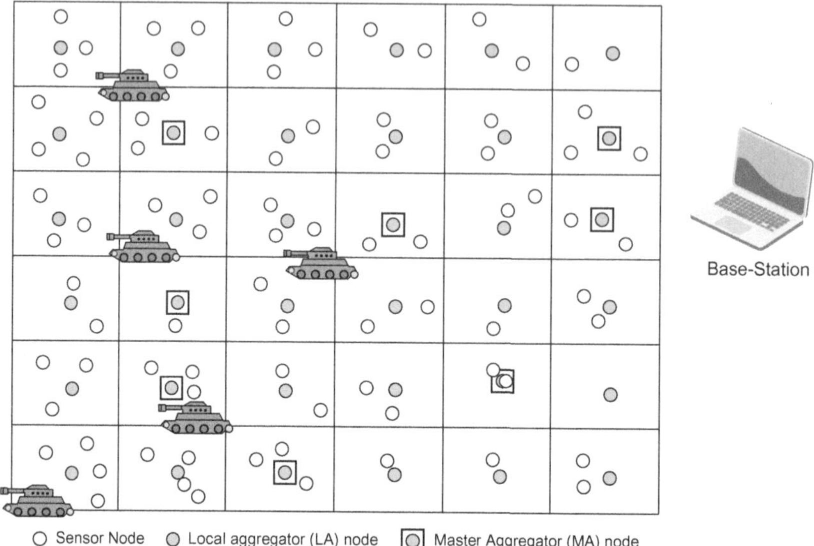

FIGURE 6.4　Regular shape tessellation applied to the network area

by several efficient, nearly optimal heuristic algorithms [6]. These heuristics draw from diverse methodologies, including genetic algorithms, k-means clustering, and greedy strategies. Additionally, the clustering-based aggregation heuristic (CBAH), detailed in further studies, aims to significantly reduce energy usage across the network, thereby enhancing its operational longevity [9]. The goal shared by these algorithms is to strategically select a subset of MAs from the pool of LAs, with the intent to maximize network durability.

In the scenarios considered in [6], LAs are envisioned to form groups that may overlap, with each group's members detecting similar environmental conditions, leading to data correlation. LAs located within these intersecting zones are responsible for transmitting their collected data to the MAs designated for each group they are part of. The CBAH approach tackles the MA selection challenge by drawing parallels to the classic bin-packing problem, differentiating itself by the complexity added due to the unknowns regarding MA identities and their respective energy consumptions. CBAH approaches this by sequentially allocating LAs to MAs, akin to filling bins of certain capacities. This suite of exact and heuristic solutions demonstrates adaptability and efficiency, showing promise for application in extensive sensor networks by delivering results that closely align with the theoretical ideal [6, 8, 9]. These methodologies underscore the potential for significant improvements in network lifetime and resource management through strategic data aggregation and routing practices.

6.2.1.3 Location-Based Routing Protocols

This routing method leverages geographical positions to identify sensor nodes, estimating the proximity of neighbouring nodes through the analysis of incoming signal strengths. By exchanging this signal strength data, nodes can infer the relative positions of their neighbours. Additionally, for more precise location information, nodes equipped with low-power GPS receivers can directly determine their locations through satellite communication via GPS technology. To enhance energy efficiency, certain location-based routing strategies incorporate mechanisms allowing nodes to enter sleep mode during periods of inactivity, thereby conserving energy. A key focus of these strategies is to increase the number of nodes in sleep mode at any given time, further reducing energy consumption across the network. The intricacies of formulating localized schedules for sleep periods to optimize energy savings have been the subject of detailed investigation [10, 11]. This section provides an in-depth overview of various geographic or location-based routing protocols, showcasing the range of techniques employed within this category to address the unique challenges of efficient and effective data transmission in sensor networks.

The geographic adaptive fidelity (GAF) routing strategy, originally conceived for use in mobile ad hoc networks, has been proposed in [10] for implementation in sensor networks. GAF organizes the network space into a series of fixed sections, establishing a virtual grid, as shown in Figure 6.5. Within each section, nodes work together, with one node taking on the role of staying active for a certain time frame to handle monitoring duties and communication with the BS, representing its zone. GAF's approach to energy conservation involves deactivating nodes that are

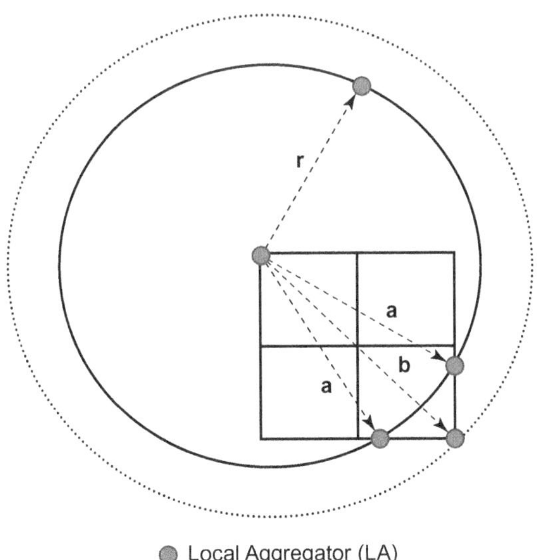

FIGURE 6.5 Fixed zoning in sensor networks

not needed for continuous operation, thereby maintaining the integrity of the routing process while conserving energy.

Nodes determine their positions within this grid using GPS technology, treating nodes within the same grid space as having equal significance regarding the cost of routing packets. This system allows for the strategic deactivation of nodes in specific grid areas to optimize energy consumption and prolong the network's operational lifespan, especially as the number of nodes increases. GAF operates through three distinct modes: discovery, active, and sleep. The active mode involves participating in routing activities, while the sleep mode turns the radio off to save energy. To manage node mobility, GAF includes mechanisms for nodes to predict when they will exit a grid area and share this information with nearby nodes. Sleeping nodes adjust their timers based on these predictions to ensure continuous routing fidelity. As the tenure of the active node concludes, sleeping nodes awaken, with one transitioning to the active state. The protocol is designed to handle scenarios with and without mobility, through GAF-basic for stationary settings and GAF-mobility for dynamic environments. The structure of the network, characterized by equal-sized square clusters, is determined by the power needed for transmission and the direction of communication. GAF's primary objective is to maintain network connectivity by always having an active node available in each grid section. Simulation studies have shown that GAF maintains performance levels comparable to those of traditional ad hoc routing protocols in terms of latency and packet delivery, all while significantly enhancing the network's lifespan through improved energy efficiency. Although GAF is a location-based protocol, its use of geographic clusters introduces a hierarchical element, with each grid's active node serving in a leadership role, like that in hierarchical protocols, albeit without engaging in data aggregation or fusion.

Geographic and energy aware routing (GEAR): Yu et al. [12] introduced the GEAR protocol, which uniquely leverages geographic information to improve the efficiency of query distribution, especially in scenarios where queries are geographically oriented. GEAR differentiates itself by incorporating energy-conscious and geographically informed strategies for neighbour selection, aiming to route packets towards a specified area in a manner that conserves energy more effectively than general directed diffusion methods.

In GEAR, each node calculates two types of costs for routing packets to their destination: an estimated cost and a learning cost. The estimated cost is calculated using the remaining energy of the node and its distance to the destination, while the learning cost is adjusted over time to better navigate around network void scenarios where no direct closer neighbour to the destination exists.

The protocol executes in two main phases: initially, it forwards packets towards the designated region using the most efficient path identified either through proximity or the learned cost. This cost is dynamically refined with each packet transmission to enhance the routing path for future packets. Once the packets arrive within the target region, they are either distributed through recursive geographic forwarding in densely populated networks, dividing the region into smaller areas for direct node-to-node forwarding, or through restricted flooding based on the specific requirements of network density. When compared to the non-energy-aware routing protocol GPSR, which also addresses routing challenges such as network voids but lacks energy efficiency optimizations and requires external location services, GEAR shows superior performance. Not only does GEAR reduce energy consumption during the routing process, but it also achieves higher rates of packet delivery. Simulation studies highlight GEAR's effectiveness, showing it can deliver between 70% to 80% more packets in scenarios with uneven traffic distribution and 25% to 35% more packets in environments with uniform traffic distribution compared to GPSR. These results emphasize GEAR's role in promoting energy savings and improving packet delivery rates in geographic-based routing strategies.

6.2.2 PATH ESTABLISHMENT-BASED ROUTING

6.2.2.1 Proactive Protocols

Proactive path establishment protocols are a category of routing protocols in WSNs that focus on establishing and maintaining routes between nodes in advance, irrespective of actual data transmission needs. These protocols maintain up-to-date routing tables, allowing for quicker data forwarding when needed. Examples include the optimized link state routing protocol. Proactive protocols are advantageous in scenarios where constant communication and low latency are critical, but they may consume more energy due to continuous route updates.

6.2.2.2 Reactive Path Establishment Protocols

Reactive routing protocols, often referred to as on-demand routing protocols, follow a distinct approach compared to their proactive counterparts. In this methodology, network paths are created solely when data transmission is required. This means that whenever a source node seeks to establish communication with a destination, it begins a process to discover a suitable route. Notable examples of such reactive protocols are the AODV and the dynamic source routing (DSR) protocols. The advantage of reactive protocols lies in their energy efficiency, as they avoid maintaining unnecessary routes and only generate paths upon demand. However, this strategy can lead to increased latency, as the route discovery phase must be completed before data transmission can commence.

6.2.2.3 Hybrid Path Establishment Protocols

Hybrid path establishment protocols aim to combine the benefits of both proactive and reactive approaches. These protocols maintain a baseline of pre-established routes, like proactive protocols, but also engage in on-demand route discovery when needed. The idea is to strike a balance between maintaining updated routing information and avoiding unnecessary energy expenditure. Hybrid protocols, like the zone routing protocol (ZRP), are designed to adapt to dynamic network conditions, offering a compromise between energy efficiency and low latency. They are suitable for scenarios with varying communication patterns and changing network topologies.

6.2.3 Protocol-Oriented Routing

6.2.3.1 Multipath Routing

This section delves into the study of routing protocols that employ multiple pathways to bolster network robustness, emphasizing fault tolerance and reliability. The durability and efficiency of these protocols are evaluated by their capacity to provide alternative routes between a source and destination should the main path encounter issues. Utilizing multiple pathways can lead to higher energy usage and increased network traffic, as it necessitates periodic communications to keep these alternative routes viable. An innovative approach suggested by authors in [13] involves routing data via nodes that possess the highest remaining energy, thereby optimizing path selection dynamically as conditions change. This strategy prioritizes the use of the main path until its energy resources dip below those of an available alternate path, effectively extending the operational lifespan of nodes on the primary route. Nonetheless, the discussion in the literature does not extend to the potential costs associated with switching paths, leaving an area of the analysis open for further exploration.

The research outlined in [14] explores the utilization of a collection of suboptimal pathways to improve the longevity of network operations. These pathways are elected based on a probabilistic model that evaluates the energy expenditure associated with each route. This approach navigates the balance between reducing overall power consumption and optimizing the distribution of remaining energy across the network. Another study, [15], introduces an algorithm that slightly modifies the criteria for residual energy to prioritize paths that are more energy-efficient, thereby offering a nuanced approach to path selection. Further investigation into multipath routing, as presented in [16], aims to boost the reliability of WSNs in scenarios where network conditions are less than ideal. This method tackles the dilemma of balancing network traffic with the need for dependable data transmission by introducing a redundancy strategy that varies with the multipath degree and the likelihood of path failure. By dividing the original data packet into several subpackets and dispatching these via multiple routes, the strategy ensures data integrity even when individual subpackets fail to reach their destination. The findings indicate that exceeding an optimal level of multipath usage could inadvertently raise the overall failure risk, given a specific threshold for node failure probability.

The concept of directed diffusion is highlighted as a strong framework for implementing multipath routing strategies. Following this framework, [17] delves into a multipath routing technique that identifies partially disjoint paths, enhancing the resilience of WSNs against failures. The introduction of braided paths, which are secondary routes closely aligned with the main path in terms of cost due to their geographical and energy similarities, aims to reduce the maintenance burden associated with multiple pathways. This method underscores the potential for energy-efficient failure recovery within sensor networks, leveraging the strategic use of alternate pathways to ensure continuous network operation.

6.2.3.2 Query-Based Routing

This routing method is characterized by destination nodes broadcasting requests for specific information (or sensing tasks) across the network. Nodes that have the requested data then reply directly to the query initiator. These queries are typically formulated in natural language or through sophisticated query languages, making the process intuitive and flexible. A prime example of this approach is seen in the directed diffusion routing protocol. In directed diffusion, the BS disseminates interest messages throughout the network, initiating the creation of data flow gradients from data sources back to the BS as the request spreads. When a node with data matching the BS's request is identified, it transmits the information back along these predefined gradients. Throughout this process, data consolidation techniques, such as eliminating redundant data, are employed to reduce energy expenditure during transmission, enhancing the network's efficiency.

The rumour routing protocol [18] introduces a novel approach to establishing directed paths towards specific events within a network by deploying long-lasting agents. These agents are responsible for creating pathways towards new events they encounter and refining existing paths to enhance efficiency or shorten routes. Within this protocol, each network node keeps a record of its immediate connections and an events table, which is periodically updated as new events are detected. Nodes can spontaneously generate these agents according to a probabilistic model. Each agent is endowed with an events table, which gets synchronized with the nodes it visits, and it operates within a predefined lifespan limited by the number of hops it can make. This mechanism encourages nodes to hold off on initiating queries until they ascertain a pathway to the sought-after event. In instances where a direct route to an event isn't known, a node may send out a query in a randomly chosen direction, awaiting a reply. Should a response not materialize within a given time frame, the node then resorts to broadcasting the query across the entire network. This protocol adeptly manages the dual challenges of responding to data requests and discovering new events within a sensor network, striking a balance between minimizing unnecessary network traffic and ensuring timely event detection and data retrieval.

6.2.3.3 QoS-Based Routing

Routing protocols within WSNs that prioritize quality of service (QoS) face the intricate task of striking a balance between conserving energy and maintaining high data quality. These protocols must navigate the complexities of achieving specific QoS benchmarks, which encompass metrics such as latency, energy efficiency, and bandwidth, to ensure the reliable and effective delivery of data to the BS. The sequential assignment routing (SAR) [19] protocol is a pioneering approach in QoS-aware routing within WSNs. SAR integrates several critical factors in its routing algorithm: the energy levels of nodes, the quality of service on potential paths, and the priority assigned to each data packet. By adopting a multipath strategy and employing local schemes for path restoration, SAR constructs a routing tree that connects source to destination nodes while circumventing those with insufficient energy or subpar QoS assurances.

In its operation, SAR computes a QoS metric that combines additive QoS properties with a weighting factor that reflects the priority of the data packet. The objective of SAR is to reduce this weighted QoS metric across the network for the duration of its lifetime. The BS periodically triggers a re-calculation of paths to adapt to any changes in the network topology. In addition, SAR features a localized path restoration mechanism that contributes to the network's resilience and recovery from failures. Simulations have shown that SAR is more power-efficient than algorithms focusing solely on minimizing energy consumption, although it does require significant overhead for maintaining routing tables and states, which becomes more challenging in larger networks.

Another QoS-centric protocol, known as SPEED [20], has been developed to provide soft real-time, end-to-end delivery guarantees. Leveraging geographic forwarding, SPEED's objective is to maintain a predefined delivery speed for each packet, which allows for the estimation of end-to-end delays based on the distance to the BS. Furthermore, SPEED incorporates mechanisms to manage network congestion effectively. The routing decisions in SPEED are powered by the stateless geographic non-deterministic forwarding (SNGF) module, which evaluates delays by timing the acknowledgements (ACKs) received. SNGF selects forwarding nodes that can meet the set speed criteria or adjusts the delivery speed based on the observed relay ratios. Comparative studies indicate that SPEED offers significant improvements over protocols such as dynamic source routing (DSR) and AODV routing in terms of end-to-end delay and packet delivery ratios. Moreover, SPEED has shown to be more energy-efficient due to its streamlined routing process. Nonetheless, SPEED's primary focus is not on energy metrics, suggesting that further analysis alongside an energy-aware routing protocol is essential to gain a fuller understanding of its performance in energy consumption.

6.2.3.4 Coherent and Non-Coherent Processing

Data processing is a pivotal component in the functionality of WSNs, with routing methodologies making use of different data processing strategies. Commonly utilized are two distinct approaches: coherent and non-coherent data processing-based routing. In non-coherent data processing-based routing, sensor nodes undertake preliminary processing of the raw data. This local processing could include filtering noise, compressing data, or extracting relevant features before the data is sent to aggregator nodes for further refinement. This method emphasizes reducing the volume of data required for transmission, thus aiming to conserve energy. Conversely, coherent data processing-based routing is characterized by sending data with only basic processing performed at the sensor nodes. The primary functions at this stage include adding time stamps, identifying, and eliminating duplicates, and other simple tasks that do not significantly alter the data volume. The goal here is to enable energy savings by minimizing the processing load on individual sensors and utilizing the more capable aggregators for intensive data processing tasks.

Non-coherent data processing within wireless sensor networks is a structured approach that unfolds across three distinct stages. Initially, the process begins with target detection, data gathering, and preliminary data refinement. In this first phase, sensor nodes identify targets and perform initial processing tasks on the collected data. The subsequent phase involves nodes opting into collaborative roles within the network, a stage known as membership declaration. The final stage is central to the network's efficiency: the election of a central node. This pivotal node is chosen based on its available energy and its ability to perform complex computational tasks.

In [19], various algorithms are put forward for managing non-coherent processing. One such proposal is the single winner algorithm (SWE), which selects an individual aggregator node to take the lead, prioritizing nodes with higher energy levels and superior processing power. An elaboration on this concept is the multiple winner algorithm (MWE), which is designed to enhance energy efficiency by curbing the number of nodes transmitting information to the central aggregator. However, the MWE approach can introduce longer delays and increased overhead, and may also face challenges in scaling, which could affect its practical application in comparison to traditional non-coherent processing networks. In WSNs, there are hybrid protocols that blend elements from multiple categories, adding layers of complexity to the routing process. This categorization allows for a comparative analysis across a spectrum of metrics, highlighting the diverse capabilities and performance benchmarks of each routing strategy.

6.2.3.5 Metaheuristic-Based Routing

Metaheuristic algorithms are optimization techniques that draw inspiration from natural and abstract concepts to solve complex problems. Unlike exact algorithms, which aim for optimal solutions, metaheuristics prioritize speed and adaptability, making them suitable for addressing computationally challenging problems with large solution spaces. These algorithms play a vital role in various fields, including combinatorial optimization, machine learning, and operations research. In the realm of WSNs, metaheuristic algorithms are instrumental in optimizing routing protocols. WSNs comprise nodes with limited computational resources, energy constraints, and often operate in dynamic environments. Efficient data routing in WSNs is critical for conserving energy, extending network lifetime, and ensuring reliable data delivery. Metaheuristic algorithms serve several purposes in WSN routing protocols:

Energy efficiency: WSNs are energy-constrained, and metaheuristics, such as particle swarm optimization (PSO) and genetic algorithms (GA), help identify energy-efficient routes. These algorithms optimize path selection to minimize energy consumption and extend life of the nodes.

Network lifetime: Prolonging the overall network lifetime is a key objective. Metaheuristics such as ant colony optimization (ACO) and simulated annealing (SA) contribute by finding routes that distribute energy consumption more evenly across network, preventing premature node depletion.

Dynamic environments: WSNs often face dynamic and unpredictable conditions, such as changing network topologies and node failures. Metaheuristics provide adaptability to these variations, ensuring the routing protocols can adjust to dynamic environments.

Scalability: Metaheuristic algorithms are designed to handle large-scale optimization problems, making them well-suited for the considerable number of nodes in WSNs. They enable scalable solutions that can address the challenges posed by the increasing scale of sensor networks.

Genetic algorithms (GAs) have proven to be highly beneficial in addressing routing challenges within WSNs. WSNs are characterized by resource constraints, dynamic environments, and the need for energy-efficient communication. GAs provides a robust optimization approach inspired by the principles of natural selection and genetics, making them well-suited for tackling complex routing problems in WSNs.

One of the primary advantages of using genetic algorithms in WSN routing is their ability to explore a large solution space efficiently. WSNs often involve numerous nodes, each with specific energy levels, varying distances, and dynamic connectivity. GAs use a population of potential solutions encoded as individuals (chromosomes), where everyone represents a potential routing scheme. Through the iterative process of selection, crossover, and mutation, GAs evolve and refine these solutions over generations. In WSNs, the optimization goals typically include energy efficiency, network lifetime extension, and reliable data transmission. Genetic algorithms excel in finding near-optimal or optimal routes that balance these objectives. The encoding of potential routes in the chromosome allows GAs to consider multiple factors, such as energy consumption, hop count, and link quality, simultaneously. This holistic approach enables GAs to discover routing solutions that not only conserve energy but also adapt to dynamic changes in the network.

Another advantage of GAs is their adaptability to dynamic WSN environments. As network topologies change due to node failures, GAs can quickly re-evaluate and reconfigure routing solutions, ensuring resilience and continuity of data transmission. Additionally, GAs can handle the challenges posed by scalability in large-scale WSNs, providing effective solutions for networks with an increasing number of sensor nodes. Typically, routing strategies in WSNs are designed to keep routing tables compact, select the optimal path to a specified destination, and necessitate a minimal amount of messages and time to achieve network convergence [3].

6.2.4 Communication Initiator-Based Routing Protocols

There are two types of routing protocols based on the initiator of communication, namely source-initiated and destination-initiated, respectively.

6.2.4.1 Source Initiated

In a source-initiated routing protocol, the process of determining routes within a network begins with the source node, which is responsible for generating the data to be sent. This contrasts with conventional routing protocols that rely on the network's infrastructure or intermediary nodes to dictate the data transmission path. Here, the source node is granted the authority to decide on the pathway for sending its data, marking a shift towards a more source-driven approach in routing decisions. In a source-initiated routing protocol, the source node is responsible for selecting the optimal path to transmit its data towards the destination. This selection is based on various metrics or criteria, such as minimizing energy consumption, reducing latency, or avoiding congested routes. The source node begins the process of finding a route by sending out a broadcast request or query throughout the network to identify potential paths and establish a route.

This approach offers flexibility and adaptability, as the source node can dynamically respond to changes in the network environment or specific requirements of the data being transmitted. Source-initiated protocols are commonly employed in WSNs and ad hoc networks, where the decentralized nature of communication and the dynamic topology necessitate efficient and adaptive routing strategies. One well-known example of a source-initiated routing protocol is the AODV protocol. In the AODV protocol, if a source node needs to send data to a destination but lacks an existing route, it starts a route discovery phase. This is done by broadcasting a route request packet. Nodes that receive this request aid in finding a route, with the search ongoing until a direct path from the source to the destination is confirmed. The demand-driven and source-oriented characteristics of AODV make it particularly effective in environments that are both changing and limited in resources.

6.2.4.2 Destination Initiated

In a destination-initiated routing protocol, the process of discovering and establishing routes within a network is triggered by the destination node. This routing strategy contrasts with approaches where the source or intermediate nodes initiate the routing, placing the responsibility on the destination node to begin the route formation. In contrast to source-initiated protocols, where the source node takes the lead in selecting the route, destination-initiated protocols prioritize the destination node in making decisions regarding the path for data transmission.

Within a destination-initiated routing protocol framework, a node that is set to receive data or is seeking to retrieve specific information initiates the process by broadcasting a route request or enquiry across the network. This solicitation is relayed through the network's nodes until it arrives at the source node or other nodes that hold the sought-after data. Throughout this procedure, intermediary nodes contribute by relaying the request further and assisting in the formation of a pathway directed towards the destination node. The primary advantage of destination-initiated protocols lies in their suitability for scenarios where the destination has knowledge about the network conditions or specific criteria for selecting an optimal path. These protocols are often employed in WSNs and ad hoc networks, where energy efficiency, reduced latency, or other metrics influence the route selection.

A notable instance of a destination-initiated routing protocol is the dynamic source routing (DSR) protocol. Within DSR, a node looking to receive data kick-starts a route discovery phase by broadcasting a packet that requests a route, clearly indicating the intended destination. This request is passed along by intermediate nodes, continuing its journey until it encounters a node that has a route to the specified destination already mapped out. This identified route is then employed for the transmission of data that follows. Destination-initiated protocols provide adaptability and efficiency, particularly when the destination node has specific requirements or preferences for the data transmission route. These protocols are well-suited for dynamic and decentralized network environments, allowing nodes to actively participate in the route discovery process based on their specific needs.

6.3 ADVANTAGES AND LIMITATIONS OF ROUTING PROTOCOLS

WSNs are a specialized type of network used for collecting and transmitting data from a variety of remote sensors. The routing protocols play a crucial role in facilitating efficient data communication within these networks. Below are the advantages and limitations of routing protocols in WSNs.

6.3.1 ADVANTAGES OF ROUTING PROTOCOLS IN WSNs

Routing protocols in WSNs bring forth a range of advantages, gauged through parameters such as energy efficiency, scalability, fault tolerance, data aggregation, and quality of service (QoS).

a) **Energy efficiency**: Energy efficiency holds paramount importance in WSNs due to the limited energy capacity of sensor nodes (SN). Replacing or recharging batteries in deployment scenarios is challenging. Routing algorithms prioritize finding paths with the least number of hops or minimum energy consumption. Energy-efficient algorithms employ metrics considering the energy levels of sensor nodes along the route.

b) **Scalability**: WSNs, with a multitude of sensor nodes spread across diverse applications, demand scalable routing protocols. Designed to handle network growth without performance degradation, hierarchical and clustering-based routing protocols efficiently manage large WSNs. They organize nodes into clusters, reducing data routing overhead.

c) **Fault tolerance**: Deployed in harsh or remote environments where nodes can fail, WSNs require fault-tolerant routing protocols to ensure data delivery. Redundancy and multipath routing are employed to provide fault tolerance. In case of a path failure, data can be rerouted through alternative paths.

d) **Data aggregation**: WSNs generate substantial data during field monitoring. Routing protocols incorporate data-aggregation techniques, allowing multiple sensor nodes to combine and process data before forwarding it to the sink node. Data aggregation conserves energy, reduces network congestion, and optimizes data transmission.

e) **Quality of service (QoS)**: Certain applications necessitate specific QoS parameters, such as high reliability, low latency, or minimum bandwidth guarantees. QoS-aware routing protocols prioritize data packets based on application-specific requirements, ensuring timely delivery within acceptable frames.

6.3.2 Limitations of Routing Protocols in WSNs

Routing protocols encounter limitations that are evaluated through parameters such as limited resources, dynamic network topology, security concerns, scalability challenges, and application-specific requirements in WSNs.

a) **Limited resources**: Sensor nodes operate within constraints of processing power, memory, and energy. Routing protocols must navigate these limitations, creating a trade-off between resource consumption and routing efficiency. Implementing complex routing algorithms becomes challenging.

b) **Dynamic network topology**: WSNs often exhibit a constantly changing network structure caused by the movement of nodes, failures, and environmental factors. Routing protocols must adapt to these changes, with dynamic network topologies potentially leading to suboptimal routing decisions, increased overhead, and data delivery delays.

c) **Security concerns**: WSNs are susceptible to security threats such as eavesdropping, data tampering, and node compromise. Routing protocols address these concerns through secure routing protocols and encryption techniques to safeguard data confidentiality and network integrity.

d) **Scalability challenges**: While routing protocols strive for scalability, very large WSNs pose challenges in terms of control message overhead and latency. Handling numerous nodes may necessitate advanced techniques like hierarchical or geographic routing to maintain scalability.

e) **Application-specific requirements**: Different applications have varied routing requirements, and a one-size-fits-all routing protocol may not suffice. Selecting or customizing the appropriate routing protocol for specific applications poses challenges, requiring thoughtful design considerations.

In conclusion, while routing protocols in WSNs offer numerous advantages, including energy efficiency, scalability, fault tolerance, data aggregation, and support for QoS, they are not without limitations. These limitations revolve around resource constraints, dynamic network topologies, security concerns, scalability challenges, and the need to meet application-specific requirements. The selection of a routing protocol should be meticulously considered based on the specific requirements and constraints of the WSN deployment.

6.4 VARIABLE SENSING RANGE (VASERA) AND ITS IMPACT IN WSNS

The Variable Sensing Range, also known as the adjustable sensing range, addresses the issue of excessive overlapping coverage in clustering caused by using a fixed detection range [21]. Traditional deployments of sensor nodes (SNs) with fixed detection and sensing ranges result in densely populated clusters, where nodes often fall within the coverage area of multiple clusters. Figure 6.6 shows the simulation instances where circles indicate the variable sensing range of SNs. To mitigate this

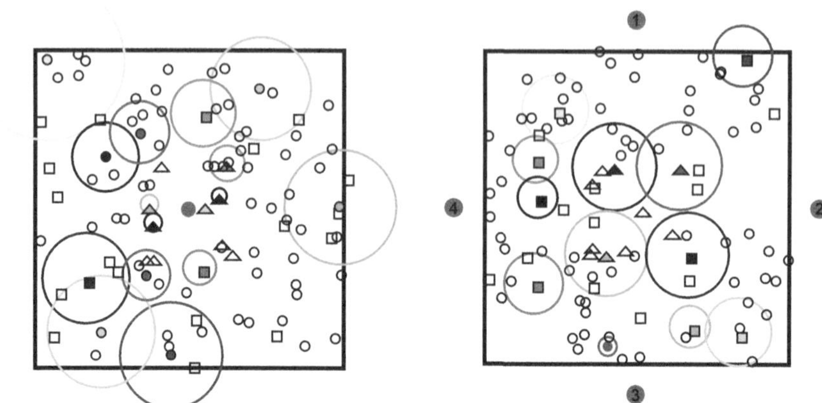

FIGURE 6.6 Simulation instances where circles indicate the variable sensing range of SNs

problem and reduce overall energy dissipation, it is helpful to restrict the sensing regions of SNs. The variable sensing range strategy efficiently manages total transmission power while preserving network linkage. Its effectiveness hinges on adapting CH sensing ranges based on their proximity to sinks. This approach enables nodes near sinks to establish direct communication with the sink itself, bypassing the need to join a cluster. This ensures complete integration of nodes into the network, even if they are not part of any specific cluster. The sensing range of an individual node is determined by its distance from the BS, while the CH's sensing range is influenced by various factors, as follows:

- **Maximum sensing range of sensor nodes**: The sensor nodes are installed with a pre-determined and unalterable maximum sensing range. This parameter represents the outer limit within which a SN can detect and communicate with other nodes or devices. This fixed range is a fundamental characteristic that influences the node's ability to perceive its environment and contribute to the overall functionality of the network.
- **The proximity of a node to the base station**: The distance between a SN and the BS is a critical factor in determining the node's operational dynamics. The proximity to the BS impacts communication efficiency, energy consumption, and data transmission speed. The nodes near the BS typically experience lower communication latency and reduced energy expenditure, contributing to the overall optimization of network performance.
- **The distance from the base station to the furthest node**: This parameter represents the maximum distance between the BS and any node within the network. Understanding the distance to the furthest node is essential for network planning and optimization. It helps in assessing the coverage area,

signal strength requirements, and potential communication challenges. Optimizing the network design based on the distance to the furthest node ensures comprehensive coverage and reliable communication across the entire network span.

This approach effectively conserves energy and extends the lifespan of a WSN. The variable sensing range addresses the hotspot issue and reduces the energy dissipation of SNs. In hotspot scenarios, nodes near the sink tend to deplete their energy quickly, while many other nodes still have enough energy. Moreover, an enormous size of clusters results in a high number of sensor nodes grouped under a single CH, leading to unnecessary communication. The adaptation of the sensing range diminishes the cluster size, eliminating uneven energy utilization in the network.

This improvement enhances energy distribution in the network and prolongs its lifetime. In this chapter, some results are mentioned by considering the variable sensing range of CHs instead of a fixed sensing range [21]. Consequently, Equation (6.2) represents the variable sensing range of a sensor node (CH_{VaseRa}) as follows:

$$CH_{\mathrm{VaseRa}} = \left(1 - \left(\frac{Dst_{max} - Dst\left(sink,\ SN_i\right)}{Dst_{max} - Dst_{min}}\right)\right) * Max_R \qquad (6.2)$$

where Max_R represents the extreme available sensing range of the SN, Dst_{max}, and Dst_{min} represent the extreme and lowest distances from the sink, and $Dst\left(sink, SN_i\right)$ denotes the distance between the sink and the ith SN. Figure 6.6 illustrates the detection range of randomly chosen nodes. From Figure 6.6, it is apparent that CHs near the sink exhibit smaller sensing ranges, while clusters located further from the sink encompass a larger cluster area, covering multiple common nodes. As a node moves away from the sink, its sensing range extends towards its Max_R value. The SNs closer to the sink have smaller $Dst\left(sink, SN_i\right)$ values, resulting in larger $Dst_{max} - Dst\left(sink, SN_i\right)$. Since the range of $(Dst_{max} - Dst_{min})$ is fixed, and its absolute value is subtracted from 1, the result becomes smaller. Consequently, CH_{VaseRa} is smaller in size for nodes near the BS. This can be readily visualized in Figure 6.6. In a single-sink scenario where the sink is located within the network area, nodes positioned along the perimeter possess the greatest sensing range. In a scenario where multiple sinks are placed near the perimeter of the area, nodes located at the central part face the greatest distance from these sinks. Consequently, these centrally located nodes are required to have an extended sensing range to maintain effective communication.

The significance of a variable or dynamic sensing range within WSNs is a pivotal factor that influences network performance, energy efficiency, and adaptability to diverse environmental conditions and application requirements. This comprehensive exploration delves into the nuanced impact of an adjustable sensing range in WSNs:

a) **Energy efficiency**: The adaptability afforded by an adjustable sensing range empowers sensor nodes to tailor their sensing capabilities to the specific demands of an application or prevailing environmental conditions. This flexibility significantly enhances energy efficiency. In scenarios where precision is not paramount, nodes can curtail their sensing range, thereby conserving energy. Conversely, when broader coverage is essential, nodes can amplify their sensing range to optimize data collection.

b) **Coverage and connectivity**: The dynamic nature of an adjustable sensing range facilitates responsive coverage and improved connectivity. Nodes, equipped with the ability to adjust their range, can strategically fill coverage gaps by expanding sensing range or fine-tuning coverage in dense areas by reducing it. This adaptability markedly improves the network's capacity to sustain connectivity and facilitate data transmission, even in challenging and dynamic environments.

c) **Data accuracy**: The impact on data accuracy hinges on the specific application. Reducing the sensing range can lead to more precise data collection in cases where nodes focus on a smaller area. Conversely, applications requiring a broader perspective may benefit from an increased sensing range, albeit with potential compromises in data accuracy.

d) **Latency and throughput**: The adjustable sensing range significantly influences latency and throughput. A reduced sensing range may alleviate latency by enabling swifter data collection from a smaller area. Conversely, an expanded sensing range may expedite latency for data travelling long distances, with potential latency increases if transmitted through more intermediate nodes. Throughput experiences a similar trade-off, balancing coverage against communication overhead.

e) **Resource utilization**: Adaptive sensing range optimization plays a pivotal role in efficient resource utilization. Sensor nodes can judiciously allocate resources by scaling their sensing range based on the network's specific needs. This adaptability empowers the network to manage power, computational resources, and memory effectively.

f) **Robustness and fault tolerance**: The adjustable sensing range bolsters the network's robustness and fault tolerance. In the event of node failure or compromise, neighbouring nodes can recalibrate their sensing ranges to cover the vacated spaces, minimizing the impact of such failures on network performance.

g) **Application-specific adaptation**: Recognizing the diverse requirements of different applications, the ability to adjust sensing ranges on a per-application basis ensures the WSN's versatility without necessitating significant hardware modifications.

h) **Security**: While an adjustable sensing range can enhance security by adapting to threats, it introduces vulnerabilities that adversaries may exploit to disrupt network operations. Implementing meticulous security measures is imperative to mitigate potential risks.

6.5 ENERGY-EFFICIENT METAHEURISTICS-BASED ROUTING PROTOCOLS

This section discusses the complete process of optimized genetic algorithms for CH selection by considering the fitness evaluation for optimized CH selection and cluster formation technique.

6.5.1 OPTIMIZED GENETIC ALGORITHMS FOR CLUSTER HEAD SELECTION

Genetic algorithms (GA) for CH selection play a pivotal role in the efficiency and performance of wireless sensor networks [22]. The GA algorithm employs evolutionary principles to strategically choose CHs that are responsible for efficient and reliable data transmission in the network. An effective utilization of the GA for the selection process of CHs helps extend the network's lifespan, enhance energy efficiency, and improve data transmission reliability.

6.5.1.1 Fitness Evaluation for Optimized CH Selection

The fitness evaluation for CH selection serves as the driving force behind enhancing the efficiency and performance of WSNs [22]. By meticulously assessing various fitness parameters, CHs are selected in such a way that maximizes energy conservation, prolongs network lifetime, and delivers reliable data transmission, motivating researchers to continuously improve and enhance these evaluation techniques. Moreover, the genetic algorithm (GA) process considers a strategic approach that repeatedly evaluates the fitness of the nodes to search for potential effective solutions. Consequently, it is crucial for this system to be sufficiently efficient in handling the calculation of all fitness criteria. If the fitness evaluation is complex, it can significantly slow down the GA strategy, negatively impacting network performance. Therefore, fitness parameters should be quickly processable with quantitative values that assess the optimality of the network arrangement. Even small changes in the parameters that are considered in the fitness evaluation may result in a conspicuous shift in the fitness value generated by the fitness function. A thoughtful selection of fitness parameters plays a crucial role in extending the stability period and enhancing energy conservation for the sensor nodes. These parameters should also consider the comparative capacity of one sensor node over another, along with the assessment of node energy for selecting CHs. Subsequently, three specific fitness parameters are considered for CH selection:

a) **Energy parameter of node**

A CH consumes a significantly higher amount of energy compared to the cluster member nodes, primarily due to its distinctive responsibilities involving data packet collection, aggregation, and forwarding towards the sink. The CH is liable to gather sensing data from each cluster node concurrently, consolidating data packets as well as to transmit them to the adjacent presented sink. Consequently, to manage the energy levels of CHs, a periodic rotation of CHs is necessary based on their remaining reserved energy. In order to extend the operational lifespan of sensor nodes, the remaining energy of a node becomes a crucial factor and is incorporated as a significant component in the fitness parameter consideration. The leftover energy of sensor nodes turn into a vital parameter in specifying the fitness function, ensuring that dynamically advanced nodes continue to lead the clusters during the reconfiguration of the network structure. The energy parameter p_E) can be calculated as represented in the following equation.

$$p_E = \sum_{k=1}^{n} \frac{N_{E\_rem}(t)}{N_{E\_max}(t)} \tag{6.3}$$

In Equation (6.3), the initial maximum energy of the kth node (N_{E_max}) indicates when the system commenced its operation, whereas N_{E_rem} signifies the leftover energy of the kth sensor node. The value of the coefficient k ranges between 1 to n by taking into account those nodes which are still active in the network. After executing the network environment, the value of N_{E_max} varies for nodes based on their specified capabilities. This parameter represents the ratio of the remaining energy to their maximum initial energies.

b) Average distance between node and sink

By considering the discussed energy model, the SNs utilize their leftover energy for transmitting the information packets to the corresponding CH or sink. To transmit the information, the distance is considered as crucial parameter. Energy usage escalates with the lengthening of the transmission gap between the sender and receiver, and inversely, diminishes as the distance decreases. The data transmission over extended distances is the most energy-intensive aspect. Therefore, incorporating a distance factor into the fitness evaluation is a significant consideration. This ensures that the CH maintains the most efficient communication path to the nearest available sink, thus promoting energy conservation. The transmission average distance parameter p_{Dst}) is expressed as follows:

$$p_{Dst} = \sum_{k=1}^{N} \left(\frac{Dst_{(n\_S)}(k)}{Dst_{F(n\_S)}(k)} + \frac{1}{Dst_{AVG(n\_S)}} \right) \tag{6.4}$$

In Equation (6.4), $Dst_{(n\_S)}$ and $Dst_{F(n\_S)}$ represent the distance between the kth sensor node with the adjacent sink and the furthest node and the adjacent sink, respectively. $Dst_{AVG(n\_S)}$ pertains to the average distance between the remaining active nodes and the designated sink. The inclusion of the average distance factor $Dst_{AVG(n\_S)}$ is intended to guide the assortment of CHs in close proximity to the sink and promote the election of CHs within a specific distance range of $Dst_{AVG(n\_S)}$.

c) Node density

This subsection emphasizes the efficient intra-group communication among the sensor nodes within a cluster, with the designated CH playing a crucial role in the well-organized management of energy resources for sensor nodes. Having a larger number of nodes operating under the control of a CH is essential for maintaining and conserving the energy reserves of cluster nodes, preventing significant depletion. Consequently, a suitable node is chosen as the CH, surrounded by a substantial number of closely interconnected SNs. This selected CH is positioned in close proximity to various nodes, resulting in a high node density within its immediate vicinity. In the event that the CH departs from the associated cluster or ceases to function, the cluster becomes fragmented, and a replacement CH is appointed from the local surroundings to fulfil the same role. This fitness criterion, which relates to node density in cluster is represented by p_{Nd} as follows:

$$p_{Nd} = \left(\frac{\sum_{t=1}^{n_C} \mathrm{Dst}_{(n-NN)}(k)}{n_C} * \frac{1}{\mathrm{Dst}_{(n-FN)}(k)} \right) \tag{6.5}$$

In Equation (6.5), $\mathrm{Dst}_{(n-NN)}(k)$ shows the distance between k^{th} SN and its nearby nodes in the cluster. $\mathrm{Dst}_{(n-FN)}(k)$ represents the distance between k^{th} node and the furthest sensor node inside the same cluster. This parameter calculates the effective distance between adjacent nodes within a cluster, with n_C denoting the total number of clusters in the network.

Network fitness function: The performance measure of a network is vital because it acts as the primary metric for evaluating the network's effectiveness and efficiency. This measure aids in identifying the most suitable CHs, contributing to improved energy savings and extended network durability. Additionally, this performance measure can be developed by integrating the previously mentioned weighted variables with certain coefficients into a unified function, referred to as the fitness function. Subsequently, in terms of the GA fitness function, Equation (6.6) represents the fitness function $\left(\mathrm{Fit}_F \right)$ to evaluate the fitness of SNs for optimize the process of CH selection.

$$\mathrm{Fit}_F = \frac{1}{\alpha * P_E + \beta * P_{Dst} + \gamma * P_{Nd}} \tag{6.6}$$

To ensure the proportionate duty of each parameter in the evaluation of fitness value calculation, α, β, and γ are considered as the weighted coefficients. These coefficients are weighted equally and considered as units, i.e. $\alpha + \beta + \gamma += 1$. The proposed GA aims to explore potential cluster structure formation, with the objective of minimizing the fitness value Fit_F throughout the evolutionary process of network chromosomes.

6.5.1.2   GA-Based Cluster Formation Technique

The genetic algorithm (GA) is a heuristic approach that follows a systematic procedure to identify optimal solutions, aiming to enhance the potential performance of network structures by simulating the principles of natural evolution. This method leverages genetic operations to construct network structures that facilitate the advancement of ideal clusters. Each new generation surpasses its predecessor in terms of quality. As the chromosomes hold crucial information about the CHs in the network, merely duplicating these chromosome structures in the offspring generation will not yield any improvements in the cluster structure. Therefore, for substantial evolutionary progress in creating ideal clusters, specific genetic operations such as crossover and mutation are essential. The combination of segments from different chromosomes, including CH formation information from various network parameters, results in the incremental development of a new network cluster structure, which can have varying degrees of effectiveness. Additionally, the most successful cluster structure obtained thus far is preserved by directly replicating it for the next generation of the population. This approach is referred to as the "elitism" methodology, which safeguards the best candidate solution achieved thus far. Moreover, it allows the chromosome to continue participating in the selection process for the development of subsequent generations. This approach ensures that the best cluster design is consistently passed on from one generation to the next, permitting adjustments based on the prevailing environmental conditions.

a) **Encoding information: genotype and phenotype**

In GA, the chromosomes store essential data about the CHs and the rest of the nodes in the network. Some chromosome structures may contain a smaller number of CHs and still effectively cover the network area, while others may have an abundance of CHs with some areas remaining uncovered. Designing a cluster structure with a higher fitness value contributes to an extended lifetime, allowing the cluster of nodes to operate for an extended duration. The genotype, which represents the sensor network's cluster structure information stored within the chromosome, serves as an automated data code that determines the selection of a sensor node as a CH or a common node. This data operation replicates the behaviour observed at the code level and is inherited across successive generations, supporting the progressive adaptation of the network. Within the outlined network architecture, it's essential for each chromosome to define "which sensor nodes are designated as CHs" and "which sensor nodes function as the regular nodes."

The genetic representation for a chromosome is illustrated in Figure 6.7, comprising a communication array that mirrors the network's node count. Each part of this array is allocated a binary value, 0 or 1. A value of 0 denotes that the associated sensor node acts either as a regular node or potentially a CH, whereas a value of 1 designates the node specifically as a CH. Therefore, the chromosome is effectively a binary string, equal in length to the network's total sensor count, with zeroes marking regular nodes and ones identifying CHs. During the setup stage, chromosomes are randomly generated, with a fundamental rule applied to cap the formation of clusters. This restriction is key to avoiding an overwhelming increase in CHs across the network space.

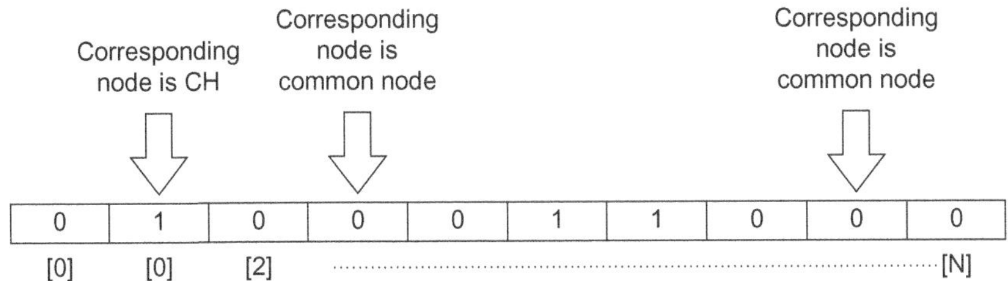

FIGURE 6.7 Genotype encoding for a chromosome

The genotypes describe the structure at a detailed level as the counterpart of phenotype. This phenotype represents a chromosome in terms of its visual or expressive attributes. Genotype refers to the genetic digital data, while phenotype is the visual or expressive representation of that data. In this study, the digital chromosome data is encapsulated in numerical variables and conveyed through colours. A sensor node is visually represented with a colour when the genotype data is set to 1 (indicating a CH), and a value of 0 signifies an empty node.

b) Crossover

The crossover operation is a genetic process that combines two chromosomes to generate a new offspring chromosome. This operation involves two parental segments and is binary in nature. However, it does not strictly limit the combination of two segments or chromosomes; instead, various types of hybrid operations, involving different segments, are described in the literature depending on the specific application they are being applied to. Here, we utilize a single-point crossover method, in which the offspring chromosome inherits half of its genetic material (pertaining to cluster structure information) from one parent chromosome, while the remaining half is derived from the other parent chromosome. This method guarantees that genetic details related to cluster configurations are transmitted to subsequent generations.

In the one-point crossover technique, a specific position, typically around the middle of the chromosome's digital sequence, is chosen as the cutting point. The resulting offspring inherits a blend of cluster data, drawing half from one parent and the remaining half from the other, thus merging two parental cluster configurations into a singular, new structure. However, it's possible for the newly formed offspring to present an issue where two CHs end up positioned too closely, leading to their disqualification based on set criteria for acceptance. Additionally, this method of crossover specifically targets active nodes for the recombination process, intentionally excluding any nodes that are inactive or non-functional. The decision to execute a crossover is governed by a predetermined probability, deciding whether to combine parent chromosomes to create offspring or to carry the parental genetic material into the subsequent generation unchanged. The notation $chromosome_R^{Pop}$ denotes the parent chromosomes considered for the crossover. As noted earlier, it's important that all genotypes maintain uniform length to ensure consistency across the genetic pool.

1. **Begin**
2. The procedure commences by randomly selecting a crossover point for executing the crossover operation on the parent chromosomes $chromosome_I^{Pop}$ and $chromosome_J^{Pop}$. Let $pointX$ be the crossover point variable.
3. $pointX \leftarrow random(0, len(chromosome_I^{Pop}))$; Choosing random trim point
4. let $child_{chromosome} \leftarrow$ new $Array(len(chromosome_I^{Pop}))$;
5. Combine digital data of Ith and J^{th} chromosome belongs to the parent.

      i.   for V $\leftarrow$ 0 to $len(chromosome_I^{Pop})$

      ii.   **If Node**$_{Population}$**(V)** is alive

      iii.   **If** V $<$ $pointX$

      iv. $child_{chromosome}$ $[V]$ = $chromosome_I^{Pop}$ (V)

             Else

      v. $child_{chromosome}$ $[V]$ = $chromosome_J^{Pop}$ (V)

         Else

      vi. $child_{chromosome}$ $[V]$ = 0
6. **If** $child_{chromosome}$ $is\ valid$

         Return best node $\leftarrow$ $child_{chromosome}$
7. **ELSE**   **Return** null
8. **End**

c) Mutation

The crossover operation inherits the characteristics of parental chromosomes and transfers them to the offspring population. The newly formed cluster structures primarily involve reconfigurations of existing ones, maintaining the status quo of common nodes and cluster nodes without significant variations. Following the concept of diversity, the evolutionary algorithm aims to avoid confinement to local optima by consistently seeking out varied cluster configurations within the network. This objective is achieved through the mutation process, which injects the required variability in CH selection. Additionally, the mutation rate is set at 1%, signifying a 1% chance of changing a digital value, thereby introducing the potential for variation and enhancing the exploration capabilities of the algorithm.

In our context, mutations primarily entail changing the value from 0 to 1 or vice versa, effectively transforming common nodes into CHs. The mutation rate must be chosen judiciously, as a higher value can lead to an excess of CHs in the sensor network or their complete removal. When converting a common node into a CH, there is a chance that the new CH may come into close proximity with an existing CH, leading to an invalid chromosome, which must be excluded from consideration. The mutation function for CH selection takes the Rth chromosome, $chromosome_R^{Pop}$, and a validity check is performed as part of the validation process. The following pseudocode presents the step by step procedure of mutation.

1. **Begin**
2. $U \leftarrow 1,2,3\ldots.len(chromosome_k{}^{Pop})$ //digital value in $chromosome_R^{Pop}$
3. **For** each U
4.   **If** $random(1) \le mutation\_rate$ **then**
5. $chromosome_R^{Pop}(U) \leftarrow !\, chromosome_R^{Pop}(U);$
6.   **If not** $Valid(chromosome_R^{Pop})$ **then**
7.   **Discard mutation**
8. **End For**
9. **End**

d) **Mating pool algorithm**

In the context of genetic algorithms and evolutionary computation, a mating pool is a group of selected individuals (chromosomes or candidate solutions) from the current generation of a population that are chosen to participate in the reproduction process. The purpose of the mating pool is to provide a set of potential parents for generating offspring in the next generation. The following pseudocode presents the step by step procedure of the mating pool process.

1. **Begin**
2. $Chromosome^{Pop} \leftarrow$ Create a pool for chromosome population
3. Initialize $Fsum \leftarrow 0$ // to hold the fitness sum of all the chromosomes in $Chromosome^{Pop}$
4. $Fsum$ $\displaystyle\sum_{q=1}^{len\left(Chromosome^{Pop}\right)} Chromosome_q^{Pop}.fitness$; Add up all wellness estimation of all chromosomes of populace.
5. Let $probability^F\,[\,]$ (empty array) .
6. **For** $p\,1\ to\ len\left(Chromosome^{Pop}\right))$ each chromosome's fitness probability value is added to $probability^F$ array i.e. $probability^F.push\,((Chromosome_q^{Pop}.fitness\,/\ Fsum\,)\,*100)$
7. Let $selection^{set} \leftarrow [\ ]$ initialize an empty selection pool array to store the indices from $chromosome^{Pop}$
8. **For** $y\,1\,to\,len\left(Chromosome^{Pop}\right)$ **repeat steps 9 to 11**
9.   calculate $m\ ceil\,(probability^F[y]\,*\,Chromosome_y{}^{Pop}.fitness)$
10.   **for** $j\,1\,to\,m$
11. $selection^{set}.push\,(y)$
12. **Return** $selection^{Set}$
13. **End**

6.5.1.3 Work Flow of the GA-Based Technique

In this section, we discuss the workflow of GA-based techniques, as shown in Figure 6.8. Initially, sensor nodes are deployed in monitoring areas along with the sinks, positioned randomly according to a specified deployment procedure. Various scenarios are considered for sink deployment, including single static and movable sinks, as well as multiple static and movable sinks. To enhance network efficiency, we also consider variable sensing ranges and direct data collection sensor capabilities. After setting up the above, the GA protocol consists of two distinct phases: the setup phase and the steady phase. During the initial setup phase, essential parameters for the genetic algorithm are established, encompassing the size of the population, rates of mutation and crossover, and the number of generations. Subsequent to this setup, random chromosome values are created and then assessed through a specific validation criterion to prevent CHs from being positioned too closely together. Node fitness is evaluated based on specific parameters to identify the best-fitted sensor node as a CH. The population undergoes evolution using genetic procedures, excluding dead nodes. Fitness-proportionate selection is implemented, an elitism strategy preserves the best node as a CH and one-point crossover, and mutation generates the offspring generation.

In the steady phase, data transmission operations are initiated based on TDMA rules. Cluster heads play a crucial role in collecting and transmitting aggregated data, with data transmission contingent on sensed values and predefined thresholds. The energy levels of sensor nodes are monitored after each transmission, and dead nodes are identified, with the protocol ensuring that the count of dead nodes remains below the total number of sensor nodes. This comprehensive GA protocol aims to optimize network performance, prolong operational lifetime, and efficiently handle data transmission and energy management within a WSN.

GA technique:

The following pseudocode presents the step by step procedure of proposed technique.

```
1. Initialize N number of sensor nodes,
            population_size ← size of chromosomes
            cr_rate, m_rate and fitness coefficients α,β,γ,ψ,
            gen_count
2. NodePopulation ← {K =1,2,3..........N}
3. Let Chromosome^Pop ← [] signify current chromosome population
   array
4. digital genotype information of each chromosome ← Randomly
   i.  for L=1 to population_size do
   ii.         Chromosome_L^Pop ← generate_chromosome(N)
5. Let sinks ← [] // Sink co-ordinates whether single or multiple
6. Compute fitness of T^th chromosome-
   i.  for T ← 1 to len(Chromosome^Pop) do
   ii.         Chromosome_T^Pop .fitness = Fitness(Chromosome_T^Pop) using
               Equation (6.6)
```

7. **while** *stopping condition is false* **do**
    i. Update the value of *gen_count*
    ii. Update the Phenotype values.
    iii. The current generation of chromosomes are Ranked
    $$Chromosome^{Fittest} \leftarrow \text{ for each } Fitness(Chromosome_i^{Pop}) >$$
    $$Fitness(Chromosome_j^{Pop}) \text{ // } i \neq j \text{ and } 1 \leq i, j \leq population\_size$$
    iv. $Fittest^{value} \leftarrow Chromosome^{Fittest}.fitness$ // Stabilizing fitness of each chromosome
    $$\text{for } j \leftarrow 1 \text{ to } len(Chromosome^{Pop})$$
    $$Chromosome_j^{Pop}.fitness =$$
    $$Chromosome_j^{Pop}.fitness / fit\_value$$
    v. Using a rank-based selection method in mating pool, chromosomes with higher fitness will be allocated more positions in the mating pool, increasing their likelihood of being chosen as parents.
    $$pool\_selection \leftarrow mating\_pool\_procedure(Chromosome^{Pop})$$
    vi. Initialize $new\_population \leftarrow [\ ]$ and $elitism \leftarrow true$
    vii. **if** $elitism == true$ **then**
    $$new\_population.push(Chromosome^{Fittest})$$
    viii. Generate a new population of offspring by combining genetic operations on the existing population of the current generation.
    while $len(new\_population) < population\_size$
      if $random(1) < cr\_rate$ **then**
        a. $chromosome_{first} \leftarrow Chromosome^{Pop}[pool_{selection}$
        $$[random(0, len(pool\_selection))]]$$
        b. $chromosome_{second} \leftarrow Chromosome^{Pop}[pool_{selection}$
        $$[random(0, len(pool\_selection))]]$$
        c. $chromosome_{child} \leftarrow Crossover(chromosome_{first}, chromosome_{second})$
        d. $chromosome_{child} \leftarrow Mutation(chromosome_{child})$
        e. $new\_population.push(chromosome_{child})$
      else
        a. $new\_population.push(Chromosome^{Pop}[offset])$
        b. $offset \leftarrow offset + 1$
    ix. $Chromosome^{Pop} = new\_population$ // Newly developed population.
    x. Update fitness value (newly $Chromosome^{Pop}$)
    $$\text{for } J \leftarrow 1 \text{ to } len(Chromosome^{Pop}) \text{ do}$$
    $$Chromosome_j^{Pop}.fitness = Fitness(Chromosome_j^{Pop})$$

8. **end while**

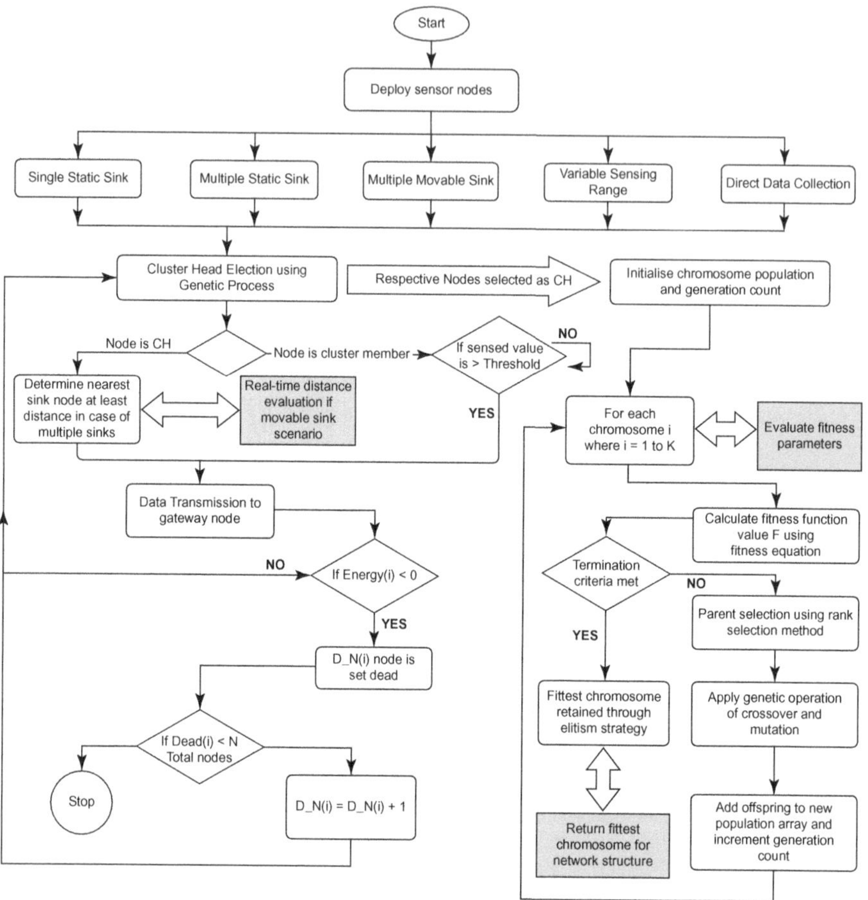

FIGURE 6.8   Flow chart of the GA-based techniques

The following outlines the description of the proposed technique in a step-by-step manner, providing a comprehensive understanding of its workings:

Step 1: Initialize the parameters of the genetic algorithm, including the number of SNs as the population size in network, mutation rate (m_rate), crossover rate (cr_rate), and the generation count (gen_count)). Additionally, assign explicit values to the weight coefficients α, β and γ for fitness parameters.

Step 2: Create a set of SNs and add them to the $Node_{Population}$ array. These sensor nodes are randomly positioned according to a proposed deployment procedure.

Step 3: Generate random chromosome values after a validation process and insert them into the $chromosome\_pool$, denoted as the $chromosome^{Pop}$

array. The chromosome population undergoes a validation rule to ensure that no two CHs are located in close proximity to each other.

Step 4: To evaluate the best-fitted node as a CH, the fitness value of nodes is calculated by evaluating the fitness parameters p_E, p_{Dst} and p_{Nd} which assess the network's performance.

Step 5: Evolve the current population of chromosomes to generate the offspring generation using genetic procedures. In this evolution process, dead nodes are not considered for the creation of the new network structure.

Step 5(i): To enable fitness-proportionate selection, a probability distribution array named *pool_selection* is constructed, taking into account the normalized fitness evaluation of each chromosome. This array ensures that chromosomes with higher fitness are given a higher probability of being selected as parents during the evolutionary process.

Step 5(ii): Employ an elitism strategy to preserve the best node as a CH, preventing the loss of the best candidate solution obtained thus far. The offspring chromosomes are continually added to the *new_population* array.

Step 5(iii): Apply one-point crossover to combine genotypes for the new generation of chromosomes. This process merges parent chromosomes with a fixed *cr_rate*, resulting in two offspring chromosomes, for example, *Crossover(ra,ra | cr_rate)=>r'a,r'b*.

Step 5(iv): To infuse diversity into the population and explore novel cluster structures, apply the mutation operator to the freshly generated offspring following the crossover operation. The mutation rate (*m_rate*), is used to determine the mutation probability. This results in mutated offspring chromosomes, e.g. *m_rate* i.e. *Mutation(ra|m_rate) = 'ra, Mutation(rb|m_rate) = 'rb*.

The protocol is divided into two separate stages: the setup phase and the continuous phase. The setup phase is dedicated to initializing the network, developing its structure, and selecting CHs.

Setup phase: During this phase, network configuration parameters and system variables are carefully established. Essential parameters for the genetic algorithm, such as the size of the chromosome population, mutation and crossover rates, and the total iterations, are specified as well. The procedures for the setup phase are elaborated as follows:

Step 1: The procedure initiates with the arrangement of nodes, which can be either homogeneous or heterogeneous in nature. For homogeneous setups, nodes are randomly positioned within the network space. Furthermore, with a single sink setup, one sink node is centrally located within the network field. Conversely, in scenarios employing multiple sink configurations, four sinks are strategically placed around the perimeter of the network area, with each one being 90 degrees away from the others.

Step 2: The process of selecting CHs involves associating chromosome genotypes with sensor nodes through the utilization of a genetic algorithm. Various network parameters and GA constraints are defined. The CH selection criteria utilized in the fitness evaluation consider factors such as remaining energy, distance optimization, node density, and comparative bias.

Steady phase: Following the establishment of cluster structures, data transmission operations are initiated in the steady phase, which is characterized by the following detailed steps:

Step 1: After completion of CH selection process, data dissemination begins from cluster member nodes based on TDMA rules. Data transmission from cluster member nodes is contingent upon the following constraint: cluster nodes can continue transmitting data if the contemporaneously measured value $\left(T\left(V\right)\right)$ exceeds the predefined threshold value $\left(J\left(V\right)\right)$; otherwise, no transmission occurs.

Step 2: When it comes to the data transmission process, CHs play a pivotal role. They collect data from the sensor nodes within their clusters and then transmit this aggregated data directly to the designated sink. In a single sink network configuration, the data is sent to the central sink. However, in a multi-sink network scenario, CHs actively search for the nearest available sink by calculating the Euclidean distance between each sink and the CH, thus ensuring efficient and timely data transfer.

Step 3: The energy level of sensor nodes is monitored after each data transmission. If it is determined that a node's left-over energy is equal to or below zero, the node is considered to be dead.

Step 4: Additionally, the protocol increments the count of dead nodes by one and proceeds with the CH selection process for the next node. It is essential to maintain a condition where the count of dead nodes remains less than the total number of sensor nodes. If, at any point, all nodes become inactive, the protocol will be terminated to avoid further operation.

6.5.2 Data Collection by Sinks from the Nearby Sensor Nodes and Cluster Heads Directly

In specific cluster configurations, forming clusters near the base station (BS) isn't always necessary. This can lead to a redundant data transmission process where data moves from a sensor node to a cluster head (CH), then to the BS, even if the node is close enough to the BS for direct communication. Implementing direct data collection by sinks minimizes this redundant data relay and unnecessary long-distance data transmission, thereby significantly improving the sensor network's overall lifespan. Figure 6.9 shows the simulation instances where sinks are deployed outside the networks and collected data directly from the SNs.

The energy consumption associated with multi-hop data transmission in the vicinity of the BS is mitigated by implementing the concept of direct data collection near the BS. As a result, the clustering process not only selects CHs but also ensures that no CH is designated in regions near the BS to promote direct data collection. This distinguishes nodes located near the BS as independent nodes for direct data transmission. Sensor nodes within a predefined range of the BS are eligible for the direct data collection process. Additionally, only nodes that are positioned in areas with poor connectivity have the capability to directly transmit data to the nearest BS (as seen in Figure 6.9). This approach effectively reduces packet traffic in the vicinity of the BS. In both figures, several areas are marked with green five-pointed stars. These stars indicate that the sensor nodes in these regions could have been designated as cluster nodes with assigned CHs, but instead, they are operating as independent nodes selected for direct data transmission.

Nodes situated in the vicinity of the BS possess the capability to send data directly to it. Nonetheless, this doesn't exclude them from being part of the clustering mechanism, where they can act as either standard nodes or CHs. As the network evolves over time, the count of operational nodes may decrease, leading to situations where forming clusters closer to the BS becomes more advantageous. Under these circumstances, the algorithm adjusts by favouring the formation of clusters near the BS. These specific areas are marked with six-pointed stars, as seen in Figure 6.9. Even though these nodes are sufficiently close to the BS to function independently, they may still be chosen as CHs or members to enhance network efficiency. This strategy of direct data forwarding markedly diminishes the overall traffic of unnecessary data packets.

6.6 PERFORMANCE ANALYSIS

This section discusses the performance investigation of the various network protocol genetic algorithm-based optimized clustering (GAOC) [22], GA-based static single

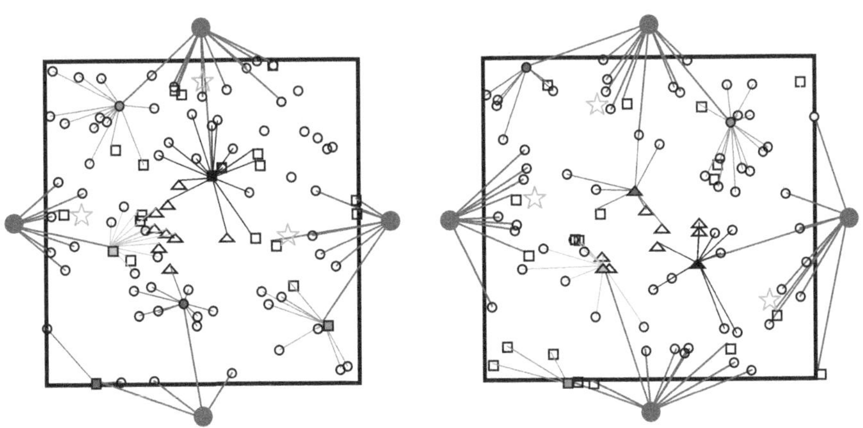

FIGURE 6.9 Simulation instances where sinks collected data directly from the SNs

sink (GenAl-StSiSi) [22], GA-based static single sink with variable sensing range (GenAl-StSi-VaseRa) [21], GA-based static single sink with direct data information collection (GenAl-StSi-DIC) [23], and GA-based static single sink with variable sensing range along with direct data information collection (GenAl-VaseRa-DIC) [24]. The evaluation of GA effectiveness is divided into three distinct sections: the use of a single static sink, multiple sinks, and mobile sinks, focusing on metrics such as network lifetime, throughput, and remaining energy. The method of data collection for each scenario, as demonstrated by current protocols, is depicted in Figures 6.10, 6.11, and 6.12. Figure 6.10 illustrates the data transmission process for a scenario with a single static sink located at the centre of the network. Figure 6.11 shows the data handling process in scenarios with multiple static sinks positioned around the network's perimeter.

Lastly, Figure 6.12 demonstrates data transmission in setups where sinks are strategically placed both inside and around the network, offering a comprehensive view of data management across different configurations.

Moreover, Figure 6.13 depicts the data transmission scenario instance of GA-based static multiple sink inside and outside with variable sensing range (GenAl-StMuSi-inout-VaseRa), by considering a multiple static sink placed both inside and outside the network with variable sensing range. Figure 6.14 presents the data transmission scenarios instance of GA-based static multiple sink outside with variable sensing

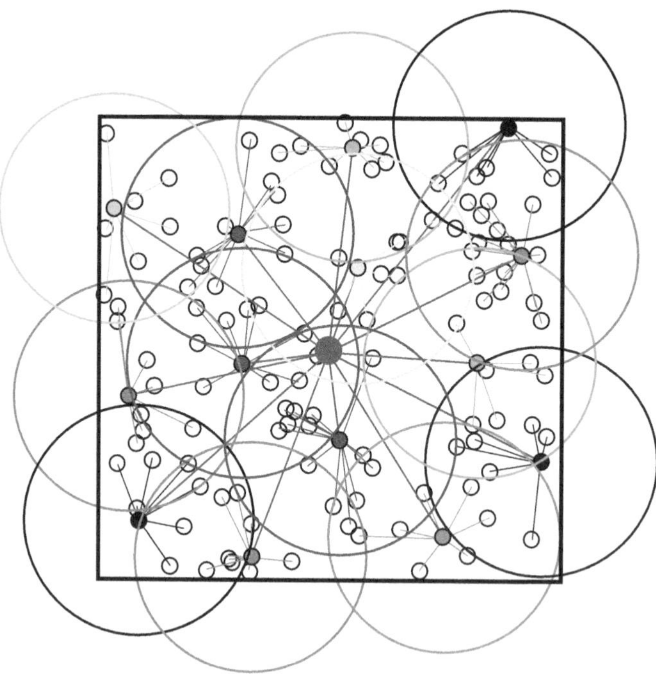

FIGURE 6.10   An instance of data communication scenario for GA-based static single sink (GenAl-StSiSi)

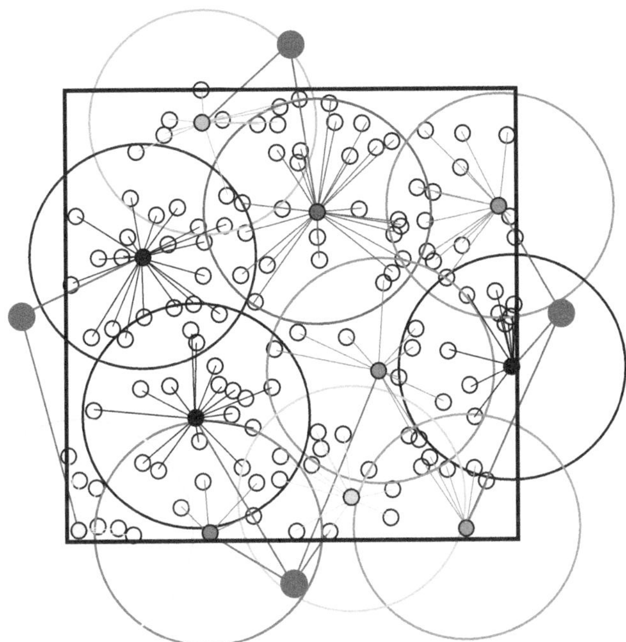

FIGURE 6.11 An instance of data communication scenario for GA-based static multiple sink (GenAl-StMuSi-out)

range along with direct data information collection (GenAl-StMuSi-VaseRa-DIC), which involve static multiple sinks outside the network with variable sensing range along with the direct data collection. In Figure 6.15, the data transmission scenario instance is depicted the data transmission scenarios of GA-based inside and outside moveable multiple sink.

6.6.1 ANALYSIS OF GA WITH STATIC SINGLE

In the forthcoming sub section, we delve into the detailed examination of GA applied to WSNs. Specifically, our focus centres on the analysis of GA performance in the context of static single and multiple sink scenarios. Through this exploration, we aim to unravel insights into the efficacy of GA-based strategies in optimizing data transmission and resource management within WSNs with varying sink configurations. The performance of existing techniques is evaluated based on the crucial metrics such as network lifetime, energy consumption, and throughput, as discussed in the following sections.

6.6.1.1 Network Lifetime

The network lifetime in WSNs is specifically emphasis on the vitality of "alive nodes" in the network. Figure 6.16 illustrates a comparative investigation of the

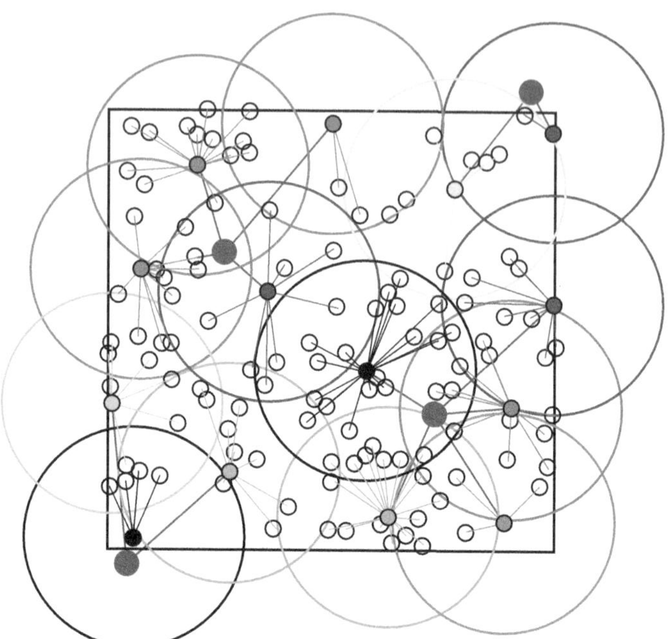

FIGURE 6.12 An instance of data communication scenario for GA-based static inout multiple sink (GenAl-StMuSi-inout)

number of alive nodes per round for the existing protocols GAOC [22], GenAl-StSiSi [22], GenAl-StSi-VaseRa [21], GenAl-DIC [23], and GenAl-VaseRa-DIC [24]. The protocol GenAl-VaseRa-DIC enhances the network lifetime by 82.88%, 80.41%, 40.70%, and 10% in comparison to GAOC, GenAl-StSiSi, GenAl-StSi-VaseRa, and GenAl-DIC, respectively. The variable sensing range along with the direct information collection strategy plays a crucial role for the extended network lifetime of GenAl-VaseRa-DIC compared to other existing techniques.

6.6.1.2 Energy Consumption

In this section, the performance of the protocols is analyzed with a specific focus on the intricate dynamics of the average residual energy of the network. This analysis aims to unravel the shades of how residual energy or the remaining energy within sensor nodes influences the overall efficiency and sustainability of WSNs. Moreover, the metric of remaining network energy is utilized to gauge the rate of energy consumption per round, as illustrated in Figure 6.17, for the protocols GAOC [22], GenAl-StSiSi [22], GenAl-StSi-VaseRa [21, 25], GenAl-DIC [23, 25, 26], and GenAl-VaseRa-DIC [24]. The protocol GenAl-VaseRa-DIC utilizes the nodes' energy efficiently in comparison to the GAOC, GenAl-StSiSi, GenAl-StSi-VaseRa, and GenAl-DIC [26]. This enhanced performance of GenAl-VaseRa-DIC is

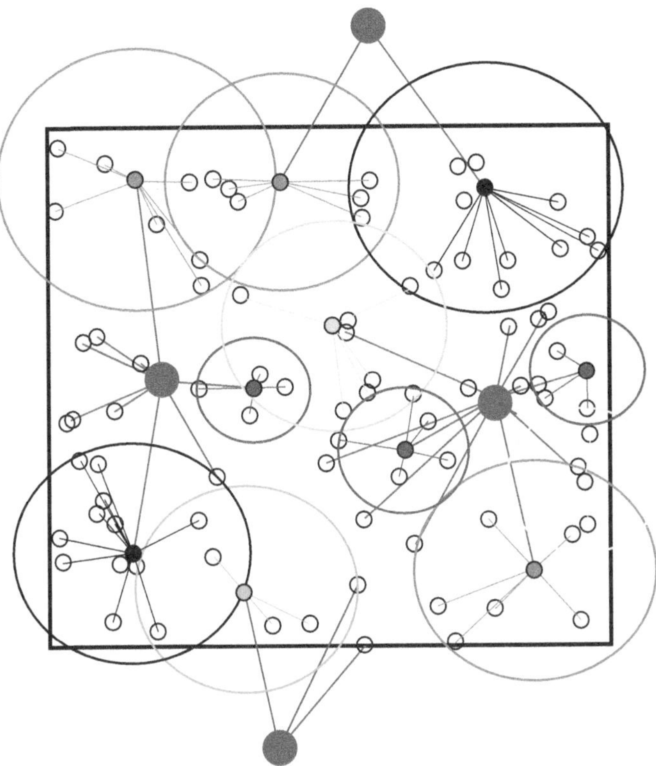

FIGURE 6.13 An instance of data communication scenario for GA-based static multiple sink inside and outside with variable sensing range (GenAl-StMuSi-inout-VaseRa)

the impact of enabling CHs to consume less energy for packet transmission, thereby conserving the overall energy reservoir of nodes.

Additionally, intra-cluster communication plays a pivotal role in preserving node energy by altering the structure of the fittest chromosome. Consequently, the genetic algorithm strategically organizes CHs based on the least distance communication paradigm.

6.6.1.3 Throughput

Throughput is a fundamental metric that helps to ensure that the protocols meets the required data transfer rate in the network. Figure 6.18 illustrates the quantity of data packets transmitted to the sink per round in homogeneous networks for the existing protocols GAOC [22], GenAl-StSiSi [22], GenAl-StSi-VaseRa [21], GenAl-DIC [23], and GenAl-VaseRa-DIC [24]. The protocol GenAl-VaseRa-DIC enhances the throughput by 53.84%, 48.20%, 41.82%, and 4.2%in comparison to the GAOC, GenAl-StSiSi, GenAl-StSi-VaseRa, and GenAl-DIC, respectively.

This enhancement is attributed to optimized CH selection considering continuous sink movement for distance optimization, thereby enhancing network durability.

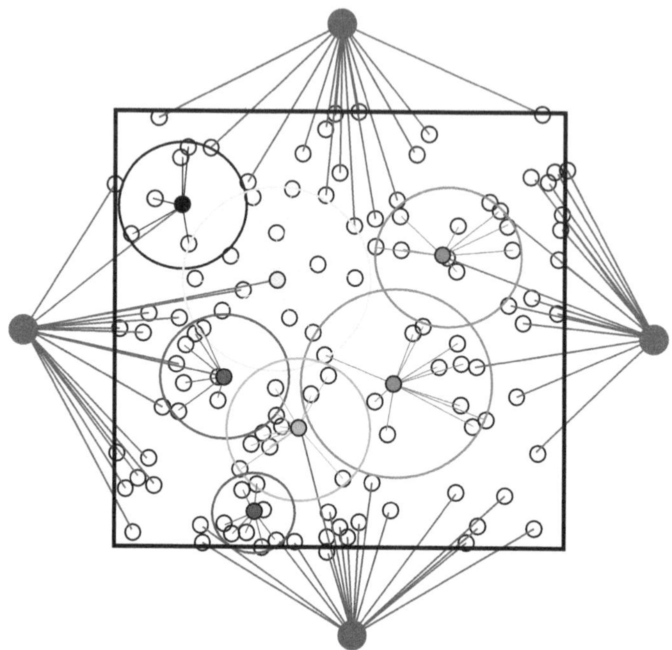

FIGURE 6.14 An instance of data communication scenario for GA-based static multiple sinks outside with variable sensing range along with direct data information collection (GenAl-StMuSi-VaseRa-DIC)

6.6.2 Analysis of GA with Multiple Sinks

In this section, a detailed analysis of the application of GA is presented in the context of WSN with a focus on scenarios involving static multiple sinks. This evaluation aims to uncover insights into the performance and efficiency of GA-based strategies when faced with the challenges posed by multiple stationary sinks in a WSN environment. By scrutinizing the intricacies of GA deployment in this specific context, the performance of existing techniques is evaluated based on the crucial metrics such as network lifetime, energy consumption, and throughput as discussed in the following sections.

6.6.2.1 Network Lifetime

Figure 6.19 shows a comparison focused on tracking the survival rate of nodes round by round, using the multi-sink-based genetic algorithm optimized clustering (MS-GAOC) [22], GA-based static multiple sink (GenAl-StMuSi) [22], GA-based static multiple sink outside (GenAl-StMuSi-out) [22], GA-based static multiple sink inside and outside (GenAl-StMuSi-inout) [22], GenAl-StMuSi-VaseRa [24], GA-based static multiple sink outside with variable sensing range (GenAl-StMuSi-out-VaseRa) [24], GenAl-StMuSi-inout-VaseRa [24, 27], and GA-based static multiple sink outside with direct information collection (GenAl-StMuSi-out-DIC) [24].

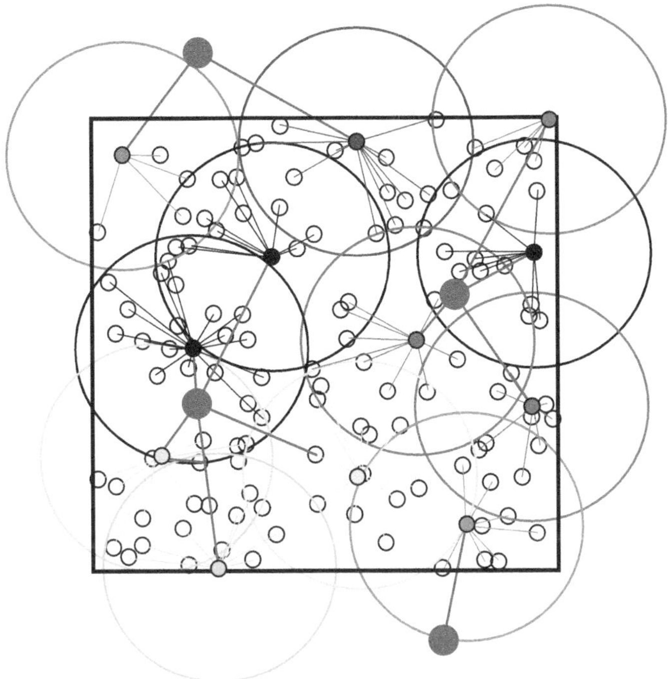

FIGURE 6.15 Scenario of data transmission in case of GA-based inside and outside moveable multiple sinks

The protocol GenAl-StMuSi-out-DIC enhances the network lifetime by 68.58%, 67.41%, 65.70%, 60.52%, 58.31%, 57.20%, and 18% in comparison to the GAOC (MS-GAOC), GenAl-StMuSi, GenAl-StMuSi-out, GenAl-StMuSi-inout, GenAl-StMuSi-VaseRa, GenAl-StMuSi-out-VaseRa, and GenAl-StMuSi-inout-VaseRa, respectively. The variable sensing range along with the direct information collection strategy plays a crucial role for the extended network lifetime of GenAl-StMuSi-out-DIC compared to other existing techniques.

6.6.2.2 Energy Consumption

This section delves into evaluating the protocols' performance, paying close attention to the nuanced aspects of the network's average remaining energy. Moreover, the metric of remaining network energy is utilized to gauge the rate of energy consumption per round, as illustrated in Figure 6.20, for the protocols MS-GAOC, GenAl-StMuSi, GenAl-StMuSi-out, GenAl-StMuSi-inout, GenAl-StMuSi-VaseRa, GenAl-StMuSi-out-VaseRa, GenAl-StMuSi-inout-VaseRa, and GenAl-StMuSi-out-DIC. The protocol GenAl-StMuSi-out-DIC utilizes the nodes' energy efficiently in comparison to the MS-GAOC, GenAl-StMuSi, GenAl-StMuSi-out, GenAl-StMuSi-inout, GenAl-StMuSi-VaseRa, GenAl-StMuSi-out-VaseRa, and GenAl-StMuSi-inout-VaseRa. This enhanced performance of GenAl-StMuSi-inout-VaseRa is the

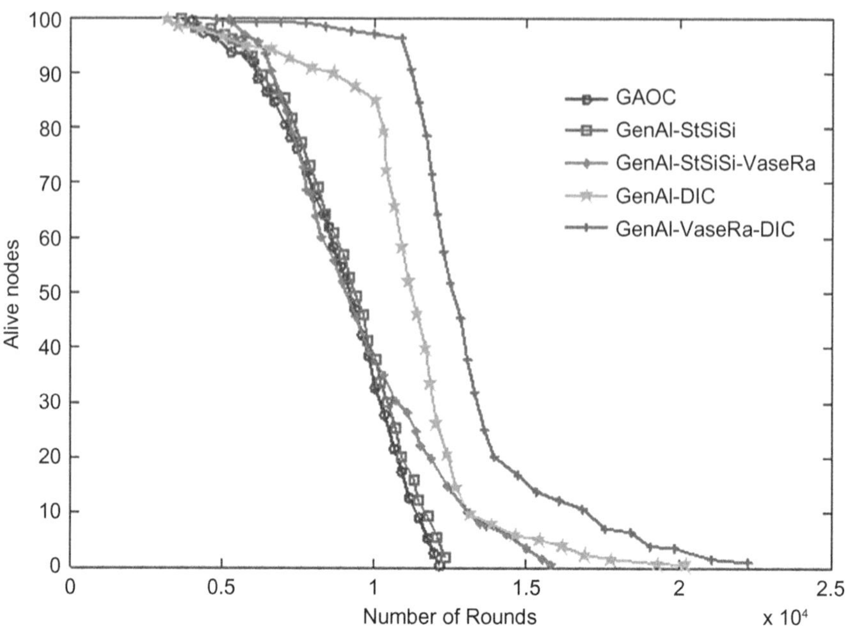

FIGURE 6.16 Alive nodes vs. number of rounds with static single sink

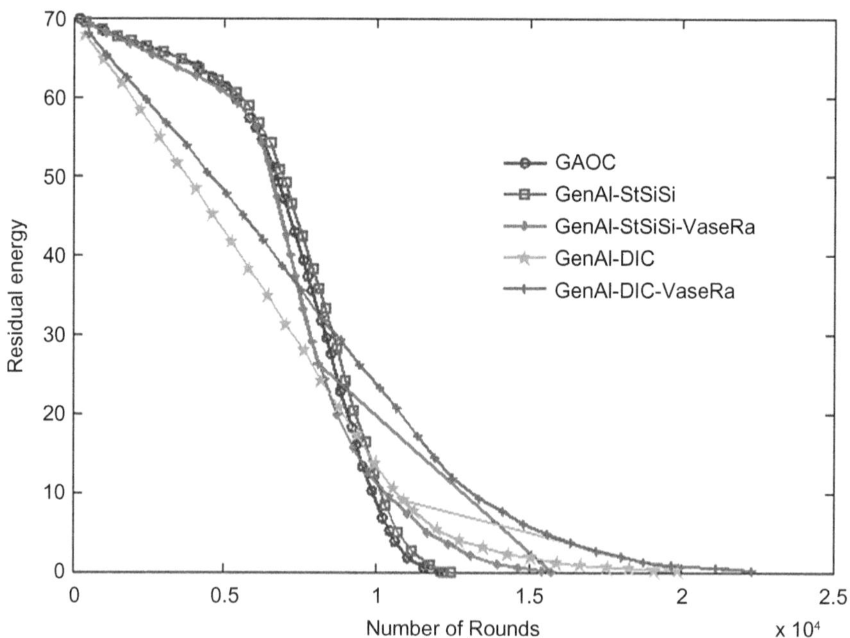

FIGURE 6.17 Residual energy vs. number of rounds with static single sink

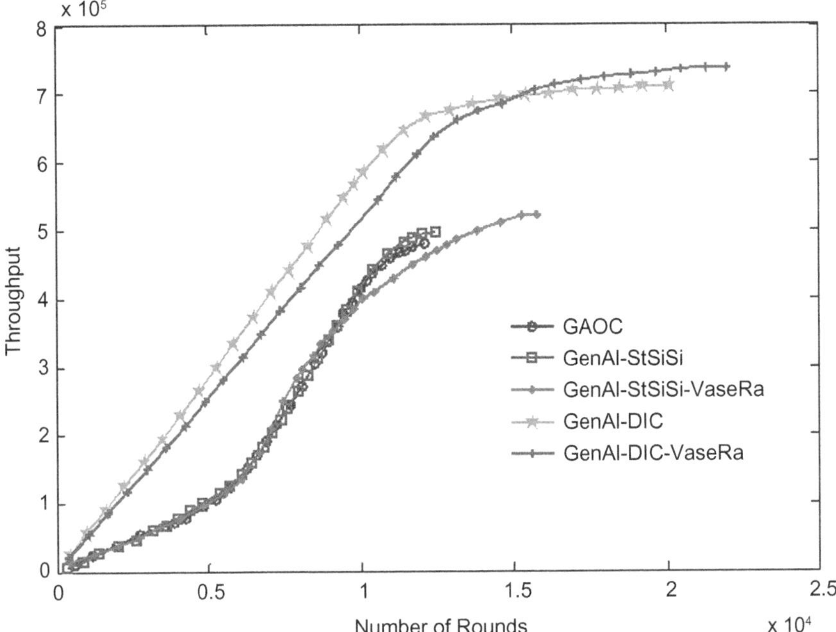

FIGURE 6.18 Throughput vs. number of rounds with static single sink

impact of enabling CHs to consume less energy for packet transmission, thereby conserving the overall energy reservoir of nodes. Additionally, intra-cluster communication plays a pivotal role in preserving node energy by altering the structure of the fittest chromosome. Consequently, the genetic algorithm strategically organizes CHs based on the least distance communication paradigm.

6.6.2.3 Throughput

Figure 6.21 illustrates the throughput with respect to the number of rounds for the existing protocols MS-GAOC, GenAl-StMuSi-outGenAl-StMuSi-inout, GenAl-StMuSi-VaseRa, GenAl-StMuSi-out-VaseRa, GenAl-StMuSi-inout-VaseRa, and GenAl-StMuSi-out-DIC. The protocol GenAl-StMuSi-out-DIC enhances the throughput by 60.48%, 47.41%, 45.70%, 45.12%, 44.43%, 37.20%, and 8.23% in comparison to the GAOC (MS-GAOC), GenAl-StMuSi, GenAl-StMuSi-out, GenAl-StMuSi-inout, GenAl-StMuSi-VaseRa, GenAl-StMuSi-out-VaseRa, and GenAl-StMuSi-inout-VaseRa, respectively. This enhancement is attributed to optimized CH selection considering energy consumption and distance optimization, thereby enhancing network durability.

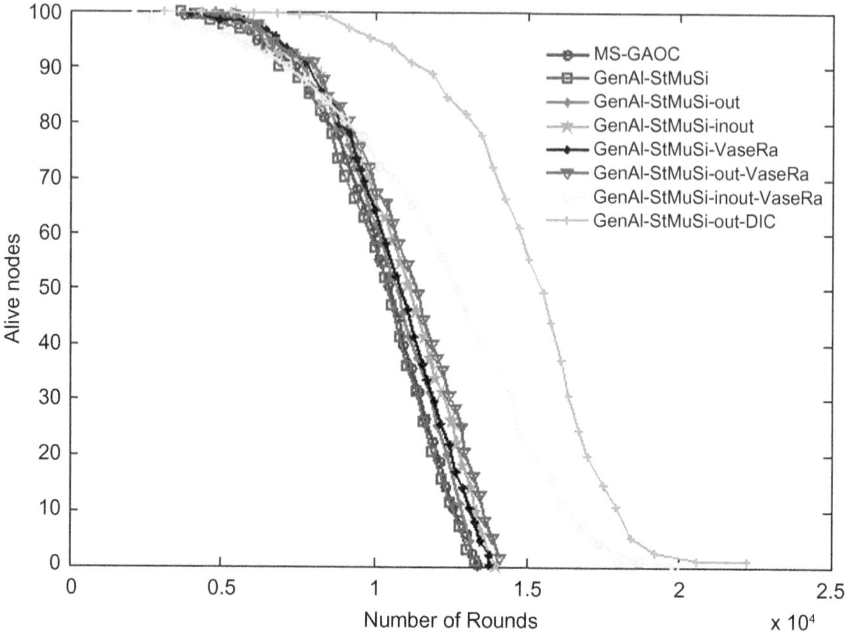

FIGURE 6.19	Alive nodes vs. number of rounds with static multiple sink

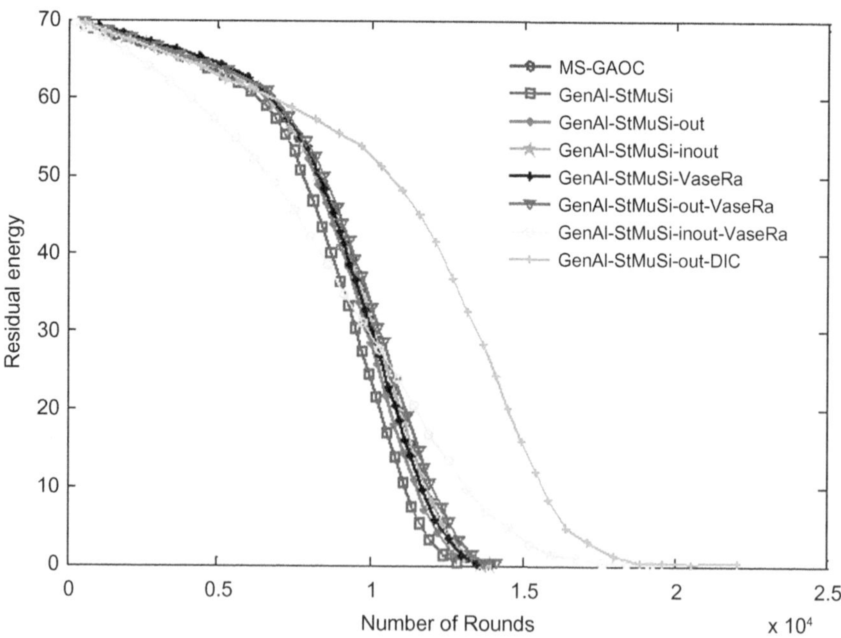

FIGURE 6.20	Residual energy vs. number of rounds with static multiple sink

6.6.3 Effect of Moveable Sink with Adjustable Sensing Range and Direct Data Collection

In this section, we delve into a comprehensive exploration of the impact of movable sinks with adjustable sensing range and direct data collection in WSN. This analysis delves into the dynamic interplay between the mobility of sinks, the adaptability of sensing ranges, and the direct collection of data. By scrutinizing the intricacies of GA deployment in this specific context, the performance of existing techniques is evaluated based on crucial metrics such as network lifetime, energy consumption, and throughput, as discussed in the following sections.

6.6.3.1 Network Lifetime

Figure 6.22 illustrates a comparative investigation of the number of alive nodes per round for the existing protocols such as GA-based moveable sink (GenAl), GA-based moveable sink with direct information collection (GenAl-DIC), GA-based moveable sink with variable sensing range (GenAl-VaseRa), GA-based outside moveable sink with direct information collection along with variable sensing range (GenAl-DIC-VaseRa-out), and GA-based inside and outside moveable sink with direct information collection along with variable sensing range (GenAl-DIC-VaseRa-inout). The protocol GenAl-DIC-VaseRa-inout enhances the network lifetime by 108.28%, 47.41%, 39.20%, and 4.56% in comparison to the GenAl, GenAl-DIC, GenAl-VaseRa, and GenAl-DIC-VaseRa-out, respectively. The variable sensing range along with the

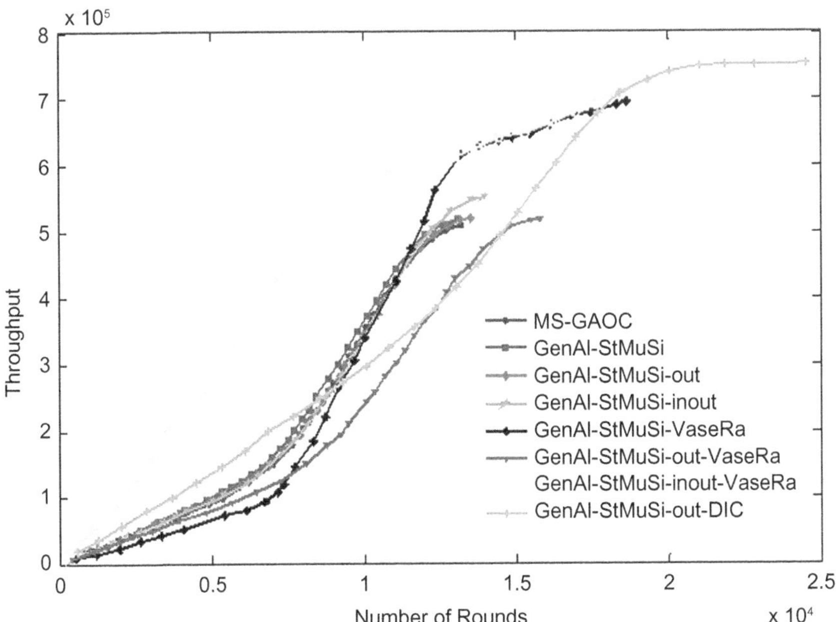

FIGURE 6.21   Throughput vs. number of rounds with static multiple sink

direct information collection strategy plays a crucial role for the extended network lifetime of GenAl-DIC-VaseRa-inout compared to other existing techniques.

6.6.3.2 Residual Energy

The performance of the protocols is analyzed with a specific focus on the intricate dynamics of the average residual energy of the network. Moreover, the metric of remaining network energy is utilized to gauge the rate of energy consumption per round, as illustrated in Figure 6.23, for the protocols such as GenAl, GenAl-DIC, GenAl-VaseRa, GenAl-DIC-VaseRa-out, and GenAl-DIC-VaseRa-inout. The protocol GenAl-DIC-VaseRa-inout utilizes the nodes' energy efficiently in comparison to the GenAl, GenAl-DIC, GenAl-VaseRa, and GenAl-DIC-VaseRa-out. This enhanced performance of GenAl-DIC-VaseRa-inout is the impact of enabling CHs to consume less energy for packet transmission, thereby conserving the overall energy reservoir of nodes.

Additionally, intra-cluster communication plays a pivotal role in preserving node energy by altering the structure of the fittest chromosome. Consequently, the genetic algorithm strategically organizes CHs based on the least distance communication paradigm.

6.6.3.3 Throughput

Figure 6.24 illustrates the throughput with respect to the number of rounds for the existing protocols such as GenAl, GenAl-DIC, GenAl-VaseRa, GenAl-DIC-VaseRa-out,

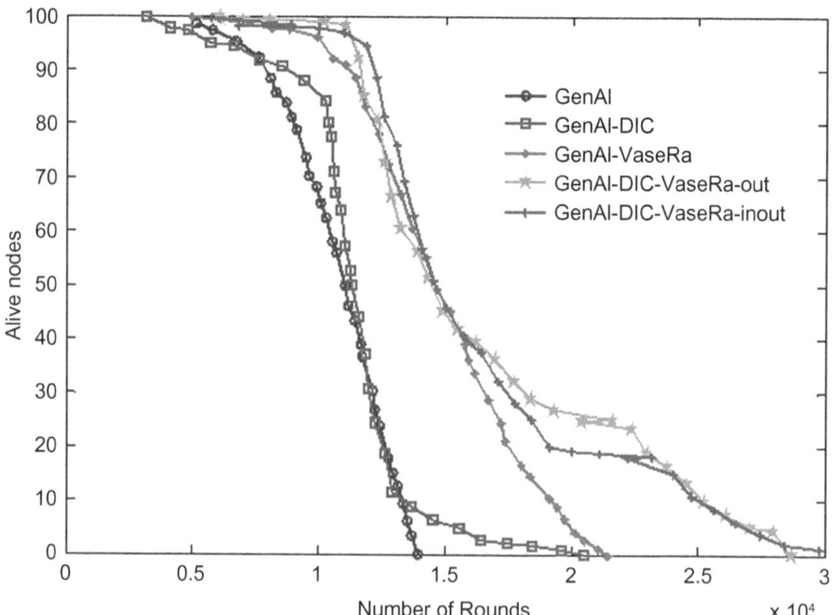

FIGURE 6.22   Alive nodes vs. number of rounds with movable sink

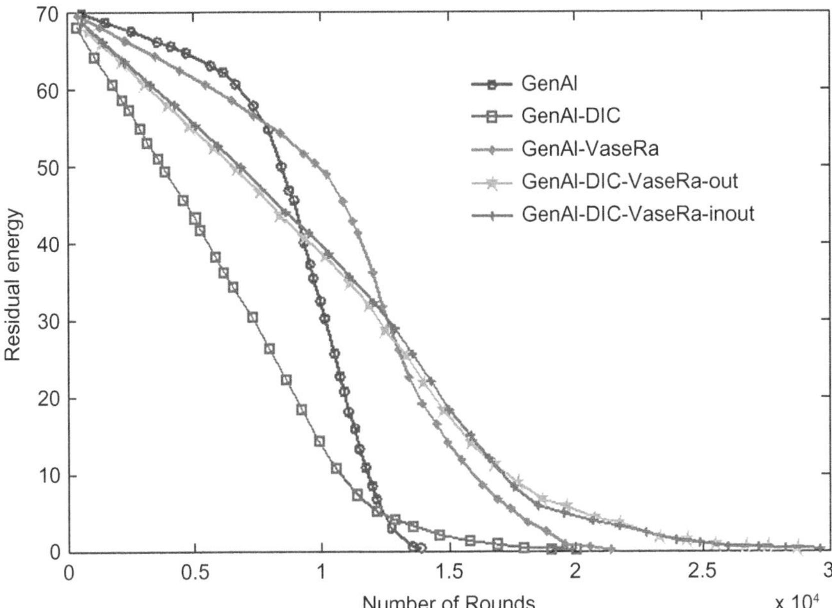

FIGURE 6.23 Residual energy vs. number of rounds with movable sink

and GenAl-DIC-VaseRa-inout. The protocol GenAl-DIC-VaseRa-inout enhances the throughput by 98.28%, 45.41%, 42.20%, and 5.56 % in comparison to the GenAl, GenAl-DIC, GenAl-VaseRa, and GenAl-DIC-VaseRa-out, respectively.

This enhancement is attributed to optimized CH selection considering continuous sink movement for distance optimization, thereby enhancing network durability. The mobility of sinks facilitates CHs to reach them most efficiently for data transmission, leading to prolonged network longevity, extended packet transmission by nodes, and a substantial improvement in throughput.

6.7 CONCLUSION

Chapter 6 embarks on an in-depth analysis of routing intricacies in wireless sensor networks, spotlighting the transformative effects of novel strategies such as modifiable sensing distances and the application of genetic algorithms for network enhancement. This section lays a robust groundwork on the myriad of routing protocols available, distinguishing them by their design, functionality, and qualitative attributes. It paves the way for a thorough discussion on the impact of dynamically altering sensing distances, a concept introduced as VaseRa, on energy usage, network durability, and information transfer efficiency. A significant segment of the chapter is devoted to examining GA-based routing protocols. The emphasis here is on fine-tuning the process of selecting cluster heads through the use of genetic algorithms that assess suitability based on essential factors such as energy status,

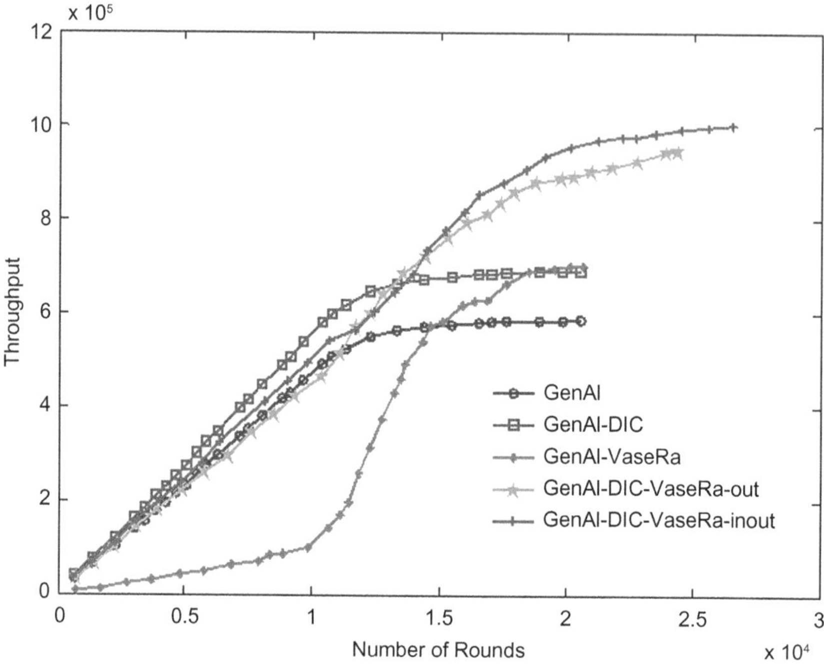

FIGURE 6.24   Alive nodes vs. number of rounds with movable sink

node distribution, and distance to the central data collector. This method underscores the capacity of GA to refine both routing and the formation of clusters, leading to marked enhancements in network operation and effectiveness.

Through an exhaustive analysis of performance, this chapter highlights the benefits of integrating GA-based tactics with modifiable sensing distances and straightforward data-gathering methods. It illustrates how these approaches surpass conventional routing protocols in various aspects – prolonging the network's active lifespan by maintaining a greater number of operational nodes over time, improving energy management to decelerate the energy depletion rate, and increasing data transfer rates to elevate the network's handling of information. Additionally, this section introduces a pioneering investigation into employing movable sinks in conjunction with the strategies mentioned above to extend the boundaries of network optimization further. This innovative method not only aims to prolong the network's functional life but also to boost its energy management and data throughput capabilities, representing a significant advancement in WSN administration and performance.

REFERENCES

1. Al-Karaki, J. N., and A. E. Kamal. "Routing Techniques in Wireless Sensor Networks: A Survey." *In IEEE Wireless Communications* 11, no. 6 (2004): 6–28. https://doi.org/10.1109/MWC.2004.1368893.
2. Kumar, A., H. Shwe, K. Wong, and P. Chong. "Location-Based Routing Protocols for Wireless Sensor Networks: A Survey." *Wireless Sensor Network* 9 (2017): 25–72. https://doi.org/10.4236/wsn.2017.91003.
3. Del-Valle-Soto, C., A. Rodríguez, and C. R. Ascencio-Piña. "A Survey of Energy-Efficient Clustering Routing Protocols for Wireless Sensor Networks Based on Metaheuristic Approaches." *Artificial Intelligence Review* 56 (2023): 9699–770. https://doi.org/10.1007/s10462-023-10402-w.
4. Intanagonwiwat, C., R. Govindan, and D. Estrin. "Directed Diffusion." Proceedings of the 6th Annual International Conference on Mobile Computing and Networking, 2000. https://doi.org/10.1145/345910.345920.
5. Heinzelman, W., J. Kulik, and H. Balakrishnan. "Adaptive Protocols for Information Dissemination in Wireless Sensor Networks." Proceeding of 5th ACM/IEEE Mobicom Conference (MobiCom '99), Seattle, WA, August 1999, pp. 174–85.
6. Al-Karaki, J. N., R. Ul-Mustafa, and A. E. Kamal. "Data Aggregation in Wireless Sensor Networks - Exact and Approximate Algorithms." Proceedings of IEEE Workshop on High Performance Switching and Routing (HPSR) 2004, Phoenix, Arizona, USA, April 18–21, 2004.
7. Xu, Y., J. Heidemann, and D. Estrin. "Geography-Informed Energy Conservation for Ad-hoc Routing." In Proceedings of the Seventh Annual ACM/IEEE International Conference on Mobile Computing and Networking, 2001, pp. 70–84.
8. Savvides, A., C.-C. Han, and M. Srivastava. "Dynamic Fine-Grained Localization in Ad-Hoc Networks of Sensors." Proceedings of the Seventh ACM Annual International Conference on Mobile Computing and Networking (MobiCom), July 2001, pp. 166–79.
9. Al-Karaki, Jamal N., and A. E. Kamal. "On the Correlated Data Gathering Problem in Wireless Sensor Networks." Proceedings of The Ninth IEEE Symposium on Computers and Communications, Alexandria, Egypt, July 2004.
10. Xu, Y., J. Heidemann, D. Estrin. "Geography-Informed Energy Conservation for Ad-hoc Routing." In Proceedings of the Seventh Annual ACM/IEEE International Conference on Mobile Computing and Networking, 2001, pp. 70–84.
11. Chen, B., K. Jamieson, H. Balakrishnan, and R. Morris. "SPAN: An Energy-Efficient Coordination Algorithm for Topology Maintenance in Ad Hoc Wireless Networks." *Wireless Networks* 8, no. 5 (September 2002): 481–94.
12. Yu, Y., D. Estrin, and R. Govindan. "Geographical and Energy-Aware Routing: A Recursive Data Dissemination Protocol for Wireless Sensor Networks." UCLA Computer Science Department Technical Report, UCLA-CSD TR-01-0023, May 2001.
13. Chang, J.-H., and L. Tassiulas. "Maximum Lifetime Routing in Wireless Sensor Networks." Proceedings of Advanced Telecommunications and Information Distribution Research Program (ATIRP2000), College Park, MD, March 2000.
14. Rahul, C., and J. Rabaey. "Energy Aware Routing for Low Energy Ad Hoc Sensor Networks." IEEE Wireless Communications and Networking Conference (WCNC), Orlando, FL, March 17–21, 2002, Vol. 1, pp. 350–5.
15. Li, Q., J. Aslam, and D. Rus. "Hierarchical Power-aware Routing in Sensor Networks." In Proceedings of the DIMACS Workshop on Pervasive Networking, May 2001.
16. Dulman, S., T. Nieberg, J. Wu, and P. Havinga. "Trade-Off between Traffic Overhead and Reliability in Multipath Routing for Wireless Sensor Networks." WCNC Workshop, New Orleans, Louisiana, USA, March 2003.

17. Ganesan, D., R. Govindan, S. Shenker, and D. Estrin. "Highly-Resilient, Energy-Efficient Multipath Routing in Wireless Sensor Networks." *ACM SIGMOBILE Mobile Computing and Communications Review* 5, no. 4 (2001): 11–25.
18. Braginsky, D., and D. Estrin. "Rumor Routing Algorithm for Sensor Networks." International Conference on Distributed Computing Systems (ICDCS'01), November 2001.
19. Sohrabi, K., and J. Pottie. "Self-Organization of a Wireless Sensor Network." *IEEE Personal Communications* 7, no. 5 (2000): 16–27.
20. He, T. et al. "SPEED: A Stateless Protocol for Real-Time Communication in Sensor Networks." In the Proceedings of International Conference on Distributed Computing Systems, Providence, RI, May 2003.
21. Nandan, A. S., Singh, S., R. Kumar, and N. Kumar. "An Optimized Genetic Algorithm for Cluster Head Election Based on Movable Sinks and Adjustable Sensing Ranges in IoT-Based HWSNs." *IEEE Internet of Things Journal* 9, no. 7 (2021): 5027–39.
22. Nandan, A. S., S. Singh, and L. K. Awasthi. "An Efficient Cluster Head Election Based on Optimized Genetic Algorithm for Movable Sinks in IoT Enabled HWSNs." *Applied Soft Computing* 107 (2021): 107318.
23. Nandan, A. S., S. Singh, A. Malik, and R. Kumar. "A Green Data Collection & Transmission Method for IoT-Based WSN in Disaster Management." *IEEE Sensors Journal* 21, no. 22 (2021): 25912–21.
24. Singh, S., A. S. Nandan, A. Malik, R. Kumar, L. K. Awasthi, and N. Kumar. "A GA-Based Sustainable and Secure Green Data Communication Method Using IoT-Enabled WSN in Healthcare." *IEEE Internet of Things Journal* 9, no. 10 (2021): 7481–90.
25. Singh, S., A. S. Nandan, G. Sikka, A. Malik, and N. Kumar. "A Genetic Algorithm Based Dynamic Transmission of Data for Communicable Disease in IoMT Environment." *IEEE Internet of Things Journal* (2023). https://doi.org/10.1109/JIOT.2023.3288614.
26. Singh, S., A. S. Nandan, G. Sikka, A. Malik, and A. Vidyarthi. "A Secure Energy-Efficient Routing Protocol for Disease Data Transmission Using IoMT." *Computers and Electrical Engineering* 101 (2022): 108113.
27. Singh, S., A. S. Nandan, A. Malik, N. Kumar, and A. Barnawi. "An Energy-Efficient Modified Metaheuristic Inspired Algorithm for Disaster Management System Using WSNs." *IEEE Sensors Journal* 21, no. 13 (2021): 15398–408.

Chapter 7

SUMMARY

The performance evaluation of wireless sensor networks (WSNs) is essential for ensuring their efficient operation and optimizing resource utilization. This chapter presents a comprehensive overview of performance metrics, analytical approaches for evaluation, and the phenomenon of self-similarity in sensor networks. Performance metrics play a crucial role in assessing the effectiveness and reliability of WSNs. The chapter examines various metrics, including jitter, latency, packet loss rate, overall throughput, frequency of retransmissions, data-aggregation error, energy consumption, and network lifetime. Each metric is discussed in detail, highlighting its significance in evaluating different aspects of network performance. The need for performance evaluation in WSNs is emphasized, considering the dynamic and resource-constrained nature of these networks. Evaluation enables researchers and practitioners to identify bottlenecks, optimize protocols, and improve overall network efficiency. Further, the analytical approaches for performance evaluation are explored, encompassing error analysis, Markov chain analysis, queuing analysis, static priority, Aloha networks, and network simulation. These approaches provide valuable insights into the behaviour of WSNs under various conditions, aiding in protocol design, optimization, and performance prediction. Also, the chapter delves into the phenomenon of self-similarity in sensor networks, wherein traffic exhibits similar statistical properties at different time scales. Self-similarity has significant implications for network performance and scalability, affecting data transmission, congestion control, and resource allocation. Understanding self-similarity phenomena is crucial for designing robust and efficient protocols capable of handling the inherent variability and unpredictability in WSNs. In conclusion, this paper provides a comprehensive framework for performance evaluation and analysis in WSNs, offering valuable insights into metrics, methods, and self-similarity phenomena. By employing appropriate evaluation techniques, researchers and practitioners can enhance the reliability, efficiency, and scalability of wireless sensor networks for diverse applications.

DOI: 10.1201/9781003427780-13

7 Performance Evaluation Metrics For Energy-Constrained Sensor Networks

7.1 PERFORMANCE METRICS

Wireless sensor networks are subsets of networks that use several sensor nodes to collect and send data from their surroundings for applications such as environmental monitoring, surveillance, healthcare, and industrial automation [1]. The performance metrics for sensor networks are designed to assess the network's efficiency, dependability, and energy usage [2, 3]. This includes jitter, network latency, packet loss rate, overall throughput, successful transmission rate, frequency of retransmissions, data-aggregation error, energy consumption, and network lifetime [4].

7.1.1 JITTER

The jitter in sensor networking signifies the variation in packet transmission delay as it traverses through the network to reach its intended destination [5]. It quantifies the irregularity or inconsistency of packet arrival timings at their destination. Jitter can be a major issue in real-time communication systems such as Voice over Internet Protocol (VoIP) conversations and video conferencing, where a continuous and predictable delay is critical for preserving communication quality [6]. Jitter is commonly created by network congestion, different routing paths, and changes in processing speeds at networking equipment along the data stream. When packet arrival times vary significantly, it can cause issues such as audio or video defects, disturbances in phone calls, and general degradation of the user experience.

Jitter is often measured in milliseconds (ms) and can be measured as interarrival, absolute, mean, and peak-to-peak jitter. Interarrival jitter refers to the time delay between consecutive packets at the receiver. Increased interarrival jitter implies instability in packet delivery. Absolute jitter refers to the absolute value of the interarrival jitter that can be expressed as its magnitude without considering its positive or negative direction. Moreover, the average of the interarrival jitter over a given time span or number of packet transmissions is referred to as the mean jitter.

DOI: 10.1201/9781003427780-14

Network administrators can also compute the difference between the maximum and minimum interarrival values of jitter for a specific time span and use it as the peak-to-peak jitter. In most network topologies, a jitter buffer is used to smooth out variations in packet arrival delays, especially during real-time communication systems. The buffer tends to momentarily store incoming packets and continuously releases them, decreasing the impact of jitter at the receiver. In a sensor network, jitter is monitored by network engineers to maintain the stability and reliability of real-time applications [7]. Excessive jitter can degrade communication quality, thereby regulating it through techniques such as traffic shaping, quality-of-service (QoS) regulations, and jitter buffers, mostly in cases where low-latency and consistent data delivery are considered crucial.

7.1.2 LATENCY

Latency refers to the total time taken for data packets initiated from the source node to reach successfully the sink or the destination node. It is a critical performance indicator in sensor networks, particularly in applications that require real-time or near-real-time data, such as industrial monitoring and automation, and healthcare. Latency in sensor networks can be influenced by various factors, including sensing, processing and transmission time, intermediate node delays, network congestion, routing algorithms, sleep cycles, wakeup mechanisms, and the underlying network topology [8, 9]. In order to minimize latency in sensor networks, there exist various techniques. Some of them are listed below:

a. Low-latency routing methods: Such routing algorithms incur low latency transmission by selecting the shortest and most reliable paths for data communication.
b. Time synchronization process: The synchronization of sensor node clocks aids in improved coordination and eliminates communication delays.
c. Data aggregation: The number of transmitted packets and delay can be reduced by aggregating data at intermediate nodes before forwarding it to the sink.
d. Energy-efficient communication mechanisms: The low-power communication protocols optimize the usage of energy, especially in transmitting data, thereby minimizing the total time taken for communication.
e. High-performance communication infrastructure: Transmission times can be reduced by utilizing sophisticated communication technology and hardware.

The permissible latency level is determined by the application requirements. Low latency is critical for real-time applications to enable fast and accurate data delivery, but non-real-time applications can endure larger latency levels.

7.1.3 Packet Loss Rate

Packet loss rate in wireless sensor network indicates the fraction of packets being transmitted from a source node but fails to reach the destined receiver or base station in the stipulated amount of time [10]. Mathematically, it is computed as the ratio of the number of lost packets and the total number of transmitted packets. It is an important performance parameter in sensor networks, as packet loss can cause data errors, decreased network efficiency, and poor application performance. Data dependability and integrity are critical in many sensor network applications, making packet loss rate a major concern. The loss of packets in sensor networks may occur due to wireless medium interference, high network congestion, queuing buffer overflows, node malfunctioning, packet collisions, transmission errors, energy constraints, etc.

In sensor networks, a low packet loss rate is often sought, especially in applications where data quality and dependability are critical [11]. To mitigate the loss of packets, various techniques are available and are described as follows:

a. Retransmission: automatic repeat request (ARQ) mechanisms can be used to initiate retransmission of lost packets.
b. Error correction codes: Forward error correction (FEC) codes can be added to the packets for recovering lost data at the receiver nodes.
c. Reliable transport protocols: Using reliable transport protocols, such as TCP (transmission control protocol), one can ensure that lost packets are retransmitted.
d. Quality of service criteria: Prioritizing packets based on their relevance and criticality can help lessen the possibility of crucial data being dropped.
e. Congestion control: Congestion control methods can be used to regulate network traffic and reduce the likelihood of buffer overflows and subsequent packet loss.

It is critical to maintain a balance between the overhead caused by reliability methods and the need for low packet loss in the sensor network as per the specific application requirements.

7.1.4 Overall Throughput

In a sensor network, overall throughput describes the total amount of data successfully communicated from source nodes to the destination or sink node in a particular time frame. Mathematically, it is expressed as ratio of total data transmitted to the time taken in transmission [12]. It is a fundamental performance statistic that assesses the network's ability to effectively transmit the content. Typically, throughput is represented in bits per second, or packets per second. The overall throughput in sensor network is influenced by significant factors, like larger packet size, rate of data transmission, channel noise, interferences from adjacent sources, limited energy and topological constraints [13]. The unit of throughput relies on the unit of data transmitted and the time unit used, hence, if data is in bits and time is measured in seconds, the throughput will be expressed in bits per second.

Achieving high overall throughput is critical in sensor networks, particularly in applications that require huge amounts of data to be transferred efficiently. High throughput ensures that the network can properly handle data generated by sensor nodes and transmit it to the destination in a timely manner. Optimizing the overall throughput in a sensor network may entail implementing efficient communication protocols, reducing packet loss, managing network congestion, and utilizing techniques such as data aggregation and compression to reduce data size and transmission time [14]. Changing the network design and node location to improve connectivity with minimized number of hops may also enhance the overall network throughput.

7.1.5 Frequency of Retransmissions

In wireless sensor networks, the frequency of retransmissions denotes the number of times data packets are retransmitted by sensor nodes owing to failed or lost transmissions [15]. When a packet is not successfully received by its intended destination, the sender node may commence retransmission attempts to guarantee that the data is eventually delivered. The frequency of retransmissions is an important feature of network performance since it directly impacts total throughput, latency, and data delivery reliability in the sensor network.

Though retransmissions are considered an effective way of improving the data reliability and transmission success rates in sensor networks, excessive retransmissions can cause communication overhead and more energy consumption. This eventually leads to degradation of network performance and may potentially cause high network congestion [16]. Therefore, to optimize the retransmissions, adaptive schemes can be used to adjust the number of repeated transmissions in a dynamic manner, in accordance with network conditions. Also, random back-off mechanisms can be used to prevent collisions and hence reduce the chances of simultaneous retransmissions from multiple nodes. Other measures such as optimal routing and application of error correction codes on corrupted packets can be used as a means to recover data, thus eradicating the need for repeated transmission cycles from multiple sources.

7.1.6 Data-Aggregation Error

In a sensor network, data-aggregation error refers to the flawed data or any other discrepancies that occurred during the process of aggregating data from different nodes in a cluster (local aggregation) or network (global aggregation). Before delivering data to the base station, data aggregation entails aggregating and summarizing data acquired by individual sensor nodes [17, 18]. In sensor networks, this technique is widely used to eliminate data redundancy, which further assists in conserving sensor power and enhancing overall network efficiency [19]. Some important factors that contribute to errors during sensor data aggregation are described as follows:

a. Loss in precision: It is possible to lose precision when combining data from different nodes, especially when averaging or summing values. As a result, the aggregated data may have minor errors when compared to the raw data.
b. Heterogeneity: The accuracy, precision, and calibration of various sensor nodes within the network may differ, resulting in differences in their measurements and influencing data aggregation.
c. Drifts in sensed data: Sensor nodes may vary in their measurements over time, resulting in inconsistent aggregated data.
d. Data communication errors: The accuracy of aggregated data may be impacted by the risk of packet loss, interference, and data corruption introduced by wireless connection.
e. Clock synchronization: If sensor nodes are not properly clock-synchronized, data aggregation may include data from various time frames, which could result in inaccuracies.
f. Faulty nodes: Erroneous data can be introduced into the aggregation process by malfunctioning or defective sensor nodes, reducing the accuracy overall.

There exist several methods that can be used in wireless sensor networks to reduce the aggregation errors in sensed data fused from multiple source nodes [20, 21].

a. Calibration in sensors: The calibration of sensor nodes on a regular basis can help increase measurement accuracy and uniformity throughout the network.
b. Data fusion techniques: In order to successfully merge data from many sources by eliminating errors, several complex data fusion techniques are used. Some data fusion techniques may be operated in a centralized manner globally by the cluster heads or sink nodes, while other techniques can be operated at the node level to aggregate samples over a time frame or to reduce redundant data transmissions.
c. Data prediction and error handling: Along with data aggregation, prediction techniques can be employed to reduce the data-aggregation rounds, thereby minimizing the chances of aggregation errors. Also, employing prediction error correction codes during data transmission can enhance the reliability of the aggregated data.
d. Quality control: During data aggregation, implementing quality control procedures can assist in identifying and handling the outlier sampled data points that may result in aggregation inaccuracies.
e. Data redundancy: Adding redundancy to data aggregation can increase data integrity and offer backup choices in the case of aggregation errors.
f. Synchronization: Assuring accurate time synchronization across sensor nodes might aid in accurately aligning data from various nodes.

In sensor networks, data aggregation involves a trade-off between data accuracy and energy efficiency. While minimizing duplicate transmissions aids in energy conservation, the aggregated data may contain some degree of mistake. The individual needs and application scenarios of the wireless sensor network should be taken into consideration when selecting data-aggregation techniques and error handling procedures.

7.1.7 ENERGY CONSUMPTION

Sensor energy is considered to be the most important resource that should be wisely used for prolonging the network lifetime [22]. This further necessitates an aggregation methodology that should use the least amount of energy. The energy consumption of a group of sensors in a particular cluster can be derived based on its communication and sleep activities, which can be further optimized [23, 24]. Initially, each sensor node is allotted energy ϵ_s, which is expressed as $\epsilon_s = \epsilon_{Active} + \epsilon_{Sleep}$. We assume all the sensors in the network are clustered, and the total number of clusters is E. Each node may belong to multiple groups, based on some similarity factor, such as positioning, energy level, or role. An ℓth similar group consists of d_ℓ nodes as its members. At the beginning, each node has an equal chance of remaining active until it gets deprived of the minimum allowable level of residual energy. This indicates that the active probability for a sensor becomes $1/d_\ell$ and the sleep probability is $1-(1/d_\ell)$. However, as the residual energy of the presently active node goes beyond the threshold, the probability for the remaining active nodes uniformly becomes $1/(d_\ell-1)$. Eventually, active probability of the member nodes that belong to the same group will increase as $1/d_\ell, 1/(d_\ell-1), 1/(d_\ell-2),...,1/3,1/2,1..$ Equation (7.1) shows the total energy usage in a group.

$$\bar{E}_{Total}(g_\ell) = \sum_{k=0}^{d_\ell-1} \left\{ \left(\frac{1}{d_\ell-k}\right)\epsilon_{Active} + \left(1-\frac{1}{d_\ell-k}\right)\epsilon_{Sleep} \right\} \tag{7.1}$$

Here, $\epsilon_{Active} = (\epsilon_{Trans} + \epsilon_{Recv})$ and ϵ_{Sleep} denote the energy required by sensor nodes in active and sleep state. The nodes in active states are engaged in transmission and reception.

$$\bar{E}_{Total}(g_\ell) = \sum_{k=0}^{d_\ell-1} \left[\left(\frac{1}{d_\ell-k}\right)(\epsilon_{Trans} + \epsilon_{Recv}) \right] + \sum_{k=0}^{d_\ell-1} \left(1-\frac{1}{d_\ell-k}\right)\epsilon_{Sleep} \tag{7.2}$$

$$\bar{E}_{Total}(g_\ell) = (\epsilon_{Trans} + \epsilon_{Recv} - \epsilon_{Sleep}) \sum_{k=0}^{d_\ell-1} \left(\frac{1}{d_\ell-k}\right) + d_\ell\epsilon_{Sleep} \tag{7.3}$$

In equations (7.2, 7.3), the term $\sum_{k=0}^{d_\ell-1}\left(\frac{1}{d_\ell-k}\right)$ is same as $\sum_{k=1}^{d_\ell}\left(\frac{1}{k}\right)$, which denotes the sum of harmonic series H_{d_ℓ}. Being a divergent series, it can be derived approximately as

$$\sum_{k=1}^{d_\ell}\left(\frac{1}{k}\right) \cong \log_e d_\ell + \gamma \tag{7.4}$$

The symbol γ is known as the Euler-Mascheroni constant. On substituting its value from Equation (7.4), the expression for total energy consumption in Equation (7.3) becomes:

$$\overline{E}_{Total}\left(g_\ell\right) \cong \left[\left(\delta+1\right)mE_{elec} + m\epsilon_{amp}\Delta^\omega - \epsilon_{Sleep}\right]\left(\log_e d_\ell + \gamma\right) + d_\ell\epsilon_{Sleep} \tag{7.5}$$

$$\overline{E}_{Total}\left(g_\ell\right) \cong \left[\left(\delta+1\right)mE_{elec} + m\epsilon_{amp}\Delta^\omega\right]\left(\log_e d_\ell + \gamma\right) + \left(d_\ell - \log_e d_\ell - \gamma\right)\epsilon_{Sleep} \tag{7.6}$$

In the above equation, E_{elec} and ϵ_{amp} refer to the energy consumed in radio electronics and amplifier circuitry. Moreover, m denotes the size of each packet in bits, and δ is the number of packets produced by individual sensors from a group during one aggregation cycle. Also, Δ is the distance between the source and destination nodes. Here, ω is denoted as the path loss component that depends upon the radio propagation model. For our derivation, the cluster sizes are considered to be large enough to sustain both the types of signal fading models, namely, free-space and multipath-fading. Accordingly, the sensor energy during transmission takes different computations owing to the threshold distance.

$$\epsilon_{Trans} = \begin{cases} m\delta E_{elec} + m\epsilon_{amp=fs}\,\Delta^{\omega=2} & \Delta < \Delta_o \\ m\delta E_{elec} + m\epsilon_{amp=mp}\,\Delta^{\omega=4} & \Delta \geq \Delta_o \end{cases} \tag{7.7}$$

Equation (7.7) represents the first order energy model where Δ_o is the threshold distance beyond which multipath fading is applied. However, free space model is used when the separation distance between the source and destination is within threshold as signal attenuation is less. Finally, in order to compute the overall consumption within a cluster C, the overall energy is computed as follows:

$$\overline{E}_C = \sum_{\ell=1}^{\overline{V}} \overline{E}_{Total}\left(g_\ell\right)$$

$$\cong \sum_{\ell=1}^{\overline{V}} \left\{\left[\left(\delta+1\right)mE_{elec} + m\epsilon_{amp}\Delta^\omega - \epsilon_{Sleep}\right]\left(\log_e d_\ell + \gamma\right) + d_\ell\epsilon_{Sleep}\right\} \tag{7.8}$$

$$\cong \left[\left(\delta+1\right)mE_{elec} + m\epsilon_{amp}\Delta^\omega - \epsilon_{Sleep}\right]\left(\log_e \prod_{\ell=1}^{\overline{V}} d_\ell + \gamma\overline{V}\right) + \psi\epsilon_{Sleep}$$

In Equation (7.8), $\sum_{\ell=1}^{\bar{\nabla}} d_\ell$ denotes the total number of member nodes in a cluster. This can be further used to compute average energy utilization across all the clusters in the network. For examining the energy consumption in a cluster for a given time period, a parameter, namely percentage of sensor energy residues $(PSER)$ is devised and is defined as follows:

$$PSER = \left(\frac{\left| E_{ini} - avg\left(\bar{E}_C\right)\right|}{E_{ini}} \bigg|_{t \in T} \right) \times 100 \tag{7.9}$$

where T is the total simulation time, and E_{ini} refers to the initial energy of a cluster which is further expressed as $E_{ini} = \sum_{s \in C} \epsilon_s \big|_{t=0}$, i.e. energy reserve at time $t = 0$.

7.1.8 NETWORK LIFETIME

The network lifetime refers to the functional lifespan of the nodes in the network and their collective capacity of performing sensor-based tasks, such as sensing, receiving, transmitting, and processing. When the first node in the network gets completely deprived of its energy reserve, the network lifetime starts declining [25]. This deduces the fact that when a node expires, it often results in connectivity gaps, thus causing disruption in networking services [26]. The remaining lifetime of sensors in a cluster C of network N is given by

$$N_C = \left(\frac{Total\,Cluster\,Members - Dead\,Nodes}{Total\,Cluster\,Members} \right) = \left(\frac{\psi - Dead\,Nodes}{\psi} \right) \tag{7.10}$$

Here, ψ refers to the total nodes grouped as a networked cluster. Due to redundant deployment of nodes, networks with a large number of sensors may frequently exhibit excess coverage. Hence, if a node belonging to a region of surplus coverage dies, then its absence may not disrupt the services. Hence, malfunctioning of such a node may not necessarily lead to network disruption and hence can be avoided in computation of network lifetime [27]. In order to preserve the longevity of a wireless sensor network and its operation, the major objective of managing network lifetime is to extend it optimally through energy-efficient protocols and methods.

7.2 NEED FOR PERFORMANCE EVALUATION

The need for performance evaluation in a sensor network is essential for a number of reasons, including the fact that it offers critical information on the behaviour, efficacy, and efficiency of the network [28]. The following are some of the main causes for which performance evaluation is necessary.

a. Network quality assurance: The performance of sensor network needs to be evaluated to ensure that it satisfies the specified dependability and quality requirements. It assists in locating potential problems, frailties, and areas that may require quality improvements in order for the network to function suitably.

b. Network optimization: Administrators can increase overall network efficiency and resource utilization by optimizing a number of parameters, including routing protocols, data-aggregation techniques, and transmission power, by analyzing the network's performance.

c. Energy constraint: Wireless sensor nodes are battery-based devices with limited energy repository. Also, sensors are often deployed in regions that are deprived of frequent human interventions, as a result of which the sensor batteries are difficult to recharge or replace. Hence, owing to severe energy constraints, it is important to track the energy performance of protocols and algorithms so that network may function properly.

d. Network reliability and robustness: The robustness and hardware reliability of the sensor network needs regular performance checks, especially under challenging circumstances such as node failures, interference, and irregular environmental changes, which are evaluated using performance analysis.

e. Throughput: Network administrators can assess the effectiveness of data transmission and the network capacity to manage data traffic in real-time or near-real-time applications by evaluating throughput. The network managers can use other performance metrics, such as frequency of retransmission attempts, network congestion, and jitter to periodically assess the network state.

f. Packet loss: Understanding data integrity and the effects of interference, congestion, and other factors on data transmission can be accomplished by analyzing packet loss and error rates. Performance analysis of communication and routing techniques are largely governed by the errors incurred in the process, which can be measured using several statistical metrics, such as mean squared, root mean squared, mean absolute, mean bias, and mean absolute percentage error metrics.

g. Network quality assessment: In order to ensure that the needs of certain applications (such as medical monitoring and industrial control) are met, performance evaluation helps to evaluate the quality assessment of the network.

h. Scalability and density: Determining the network capacity to manage an expanding number of sensor nodes and data sources without noticeably degrading performance involves evaluating its scalability. Apart from node scalability, there is an additional property of dynamic networks that includes the sparsity and density of connections between the sensor nodes. As the increasing number of sensor nodes begins to communicate, this gives rise to abundance of transmission links. Hence node scalability and link density inevitably requires performance monitoring.

i. Cost-effectiveness: In order to deploy and configure sensor nodes to attain the appropriate performance levels, performance evaluation is required for minimizing costs and assisting in decision making processes.

j. Security mechanisms: In order to adopt the proper security measures to secure sensitive data, evaluating the network's performance can assist uncover vulnerabilities and potential security threats.

k. Application specifications: Performance specifications for various applications could vary. Adjusting the network's characteristics and protocols to fit particular application requirements is made possible through performance evaluation.

l. Network planning and policies: Planning for the future can be made more effective and trustworthy by analyzing past performance heuristics of the network, especially during peak hours of communication. This helps the network administrators to plan the queuing models and buffering mechanisms to ensure minimum packet loss.

Network simulation, mathematical modelling, and practical experiments are all part of the process known as performance evaluation [29]. Overall, it largely assists network designers, researchers, and operators in making data-driven decisions. This greatly optimizes the performance of the data-oriented sensor network, thereby resulting in improved data transmission, energy efficiency, and generic network functionality.

7.3 ANALYTICAL APPROACHES FOR EVALUATION

Applications requiring particular versions of such protocol require a thorough analysis of the protocols. Markov chain analysis is one of the very basic form for analyzing several operations performed in the networking domain. For instance, error probability of the wireless channel as well as individual sources state prediction of congestion handling protocols, flow and error control, mean number of successful transmissions, etc. In addition, queuing models with finite and infinite buffer system are also extensively used for other complex-level analysis of scheduling activities. This section illustrates error analysis techniques, significance of Markov chain analysis, queuing theory, static priority and Aloha networks, and popular simulation tools used for network analysis.

7.3.1 ERROR ANALYSIS

Numerous types of errors can take place throughout the network stages, involving sensing, data transmission, and data processing of sensed data. This necessitates the study of errors while analyzing the network performance [30]. For the sensor readings to be accurate, reliable, and authenticate, the faults in sensed data must be understood and analyzed. The following are some typical faults in a wireless sensor network.

a. Sensing errors: These are the errors that typically occur during the data acquisition process performed by the individual sensors. These errors may be due to sensor calibration issues, noise, drift, or inaccuracies in the measurement.

b. Data transmission errors: These are errors that arise during the wireless transmission of data from sensor nodes to the base station or sink node. These errors can be caused by signal interference, fading, collisions, and packet loss during data transmission.

c. Data-aggregation errors: Some errors can get introduced during the process of aggregating data from multiple sensor nodes. The aggregation errors can result from inaccuracies in data fusion algorithms or loss of precision when summarizing data.

d. Data processing errors: There could be data uncertainties and ambiguities during data processing at the cluster head or base station. These errors may arise from faulty algorithms, computational errors, or issues in data analysis techniques.

e. Time synchronization errors: Inaccurate time synchronization among sensor nodes can lead to errors in data fusion and data correlation.

f. Localization errors: There could be certain errors that may occur while determining the physical location of sensor nodes, which can eventually affect the accuracy of spatially correlated sensor data.

g. Energy measurement errors: These errors may occur while estimating the energy consumption of sensor that could impact the energy-efficient protocols and operational lifetime of the network.

Error analysis in a sensor network involves identifying, quantifying, and understanding the sources of errors and their potential impact on data accuracy and network performance. It may involve the use of statistical methods, simulation studies, and experimental evaluations to assess error rates and deviations from ground truth values. By conducting error analysis, network administrators and researchers are able to:

- Identify and address sources of errors to improve data accuracy and reliability.
- Optimize sensor network protocols and algorithms to minimize impact of errors.
- Enhance energy efficiency by reducing unnecessary retransmissions caused by errors.
- Ensure that network meets desired performance requirements for specific applications.

Ultimately, error analysis is considered a crucial step in designing, operating, and maintaining the wireless sensor networks to achieve accurate and reliable data collection, communication, and processing.

7.3.2 Markov Chain Analysis

Stochastic processes (SP) play a role of eminence in modelling and evaluating the performance of wireless communication networks [31]. A stochastic process $X_S(t)$ defines the state of a system S in terms of the collection of random variables with respect to regular or continuous advancements in time t, such that $t \in T$. For instance, total trades recorded in a market M over timestamps t, subject to market volatilities can be denoted by $X_M(t)$. If the time domain is discrete or forms a countable set, the corresponding random process becomes a discrete-time stochastic process (DT-SP). A DT-SP can be mathematically represented as $\{X_S(t) \,|\, t = 0,1,2,3,...\}$. However, if the time domain is a part of uncountable set, such a process is termed a continuous-time stochastic process (CT-SP), which can be expressed as $\{X_S(t) \,|\, t \geq 0\}$. The state space of random variable refers to the set of all possible values that system S can assume. A Markovian model is a category of stochastic procedure wherein the next state of a system relies only on the current state while ignoring the past states. Statistically, a Markov process can be expressed as

$$p\left\{X_{S(n+1)} = j \,|\, X_{S(n)} = i, X_{S(n-1)} = i_{(n-1)}, X_{S(n-2)} = i_{(n-2)},..., X_{S(1)} = i_i, X_{S(0)} = i_0\right\} = p_{ij} \quad (7.11)$$

The above formula holds for the following constraints satisfied, $p_{ij} \geq 0; i, j \geq 0; \sum_{j=0}^{\infty} p_{ij} = 1 \forall i = 0,1,....$ The 1-step transition matrix, usually denoted by P, is a $N \times N$ square matrix where N is the total states that a system S can support. The matrix P basically defines the transition pattern between N system states. The Markov process can be applied for communication systems for forecasting system outcomes and estimating average functioning time.

Example 1: A wireless link modelled as Markov process, operates between two states – success (S) and failure (F) with probabilities as α and β, respectively. The probability that the link successfully functions after x time instant, provided that it currently failed, will be given by p_{ij} such that $p_{ij} \in P^x$ where i and j corresponds to F and S states.

Example 2: The transmission from two faraway situated stations A and B is accomplished by a relay station R. Let's assume that initially R spends 30% of its total time in serving station A. The probabilities for serving each station in an one-hour time span is highlighted below in the transition diagram (Figure 7.1). The estimation of average time spent by the relay station R in serving stations A and B can be computed as follows:

From the above transition diagram, the Markov transition matrix can be derived as follows:

$$P = \begin{bmatrix} 0.35 & 0.65 \\ 0.2 & 0.8 \end{bmatrix}$$

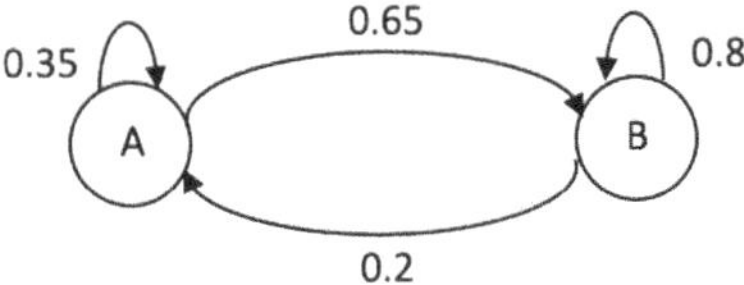

FIGURE 7.1 State transition diagram

Also, according to the initial assumption, we have $\Pi_{t=0} = \begin{bmatrix} 0.3 & 0.7 \end{bmatrix}$. Therefore, further iterations (until the limiting matrix is formed) are computed as follows:

$$\Pi_{t=1} = \Pi_{t=0} \cdot P = \begin{bmatrix} 0.245 & 0.755 \end{bmatrix}$$

$$\Pi_{t=2} = \Pi_{t=1} \cdot P = \begin{bmatrix} 0.237 & 0.763 \end{bmatrix}$$

$$\Pi_{t=3} = \Pi_{t=2} \cdot P = \begin{bmatrix} 0.236 & 0.764 \end{bmatrix}$$

$$\Pi_{t=4} = \Pi_{t=3} \cdot P = \begin{bmatrix} 0.235 & 0.765 \end{bmatrix}$$

$$\Pi_{t=5} = \Pi_{t=4} \cdot P = \begin{bmatrix} 0.235 & 0.765 \end{bmatrix}$$

Since $\Pi_{t=5} = \Pi_{t=4}$, further iterations are declined, and $\Pi_{t=5}$ becomes the limiting vector, i.e. $\Pi_{t=5} = \Pi^{*}$. Therefore, on average the relay unit serves station A with probability 0.235 and station B with probability 0.765.

7.3.3 Queuing Analysis

Queuing theory emanated from the traditional concept of queues of people waiting to avail certain services on a first-come-first-served basis. This generic theory has become basis for performance analysis of numerous applications, especially in the domain of network and computing systems [32]. This section begins with some basic queuing parameters, followed by certain models and their usage for analyzing the working of communication protocols. For instance, number of packets waiting in the buffer queue of wireless nodes to get transmitted depicts the state of a communication system.

7.3.3.1 Queuing Parameters

Some significant parameters to comprehend and analyze the performance of wireless systems, which are demonstrated using the queuing model, include throughput, average input and output traffic, efficiency, and traffic loss probability.

(a) Queue throughput: This is a queuing parameter to estimate the average rate of packets successfully transmitted per time step. In other words, queue throughput Q_{th} (also called transmission throughput from a host node) can be defined as average output traffic $\left(A_{out}\right)$.

(b) Transmission efficiency: This queuing metric can be applied to compute the fraction of packets transmitted by the host node, relative to the total packets initially scheduled to get transferred to some destined location. This parameter actually measures the efficiency of the packet queue to process the buffered transmission requests within a stipulated time frame. This essentially becomes useful to monitor the queuing performance at relay nodes or more particularly gateway nodes. This is because gateway node acts as a communication bridge between separate networked communities. Mathematically, efficiency $\left(T_{eff}\right)$ can be defined as the fraction of average output and input traffic. Ideally, the ratio should compute to 1; however due to hardware malfunctioning and processing delays, efficiency turns out be in less than 100%, i.e. $T_{eff} = \dfrac{A_{out}}{A_{in}} = \dfrac{Q_{th}}{A_{in}} < 1$. Here, A_{in} denotes average input traffic. This can be expressed as:

$$A_{in} = \left(N_a \times p_a\right) + \left(N_f \times p_f\right) \tag{7.12}$$

In the above equation, N_a denotes the number of packets arrived, and N_f is the packets that failed to arrive within a time frame. Relatively, p_a and p_f are the arrival and failed to arrive probabilities, respectively. Generally, $p_f = 1 - p_a$.

Example 3: If a host node's buffer queue is capable of transmitting five packets per seconds on average, while eight packets could arrive for transmission, then throughput $Q_{th} = A_{out} = 5$ and transmission efficiency $T_{eff} = \dfrac{Q_{th}}{A_{in}} = 0.625$. Hence, the efficiency becomes 62.5%.

(c) Traffic loss probability: This parameter is yet another significant metric that detects the probability of traffic loss. In wireless communication, when the buffer queue of the host node reaches a steady state, incoming packets either get processed by successfully moving through the queue to get transmitted, or get lost. A packet might get lost during origination or relay for several reasons. During origination, packet losses happen because of time-outs, while packets get lost in relays often due to late arrival, processing delay, network congestion, or the incapability of buffer queue (of relay node) to accommodate any further transmission requests (buffer full/corrupt). Therefore, according to network traffic conservation rule:

$$A_{in} = A_{out} + A_{lost} \tag{7.13}$$

Here, A_{lost} is average packet loss per unit time. The conservation rule is applicable only when the system is in steady state. On dividing the above equation by A_{in}, we get:

$$\frac{A_{out}}{A_{in}} + \frac{A_{lost}}{A_{in}} = 1 \Rightarrow L = 1 - T_{eff} \tag{7.14}$$

L signifies the traffic loss probability. The communication systems modelled with queues having higher transmission efficiency bear lower traffic loss.

7.3.3.2 Queuing Models

This section highlights some basic and derived queuing models that are normally applied to wireless communication. In order to describe and identify the queuing model adapted by a host node, Kendall's notation is considered quite important. D.G. Kendall prescribed a standard system that originally consists of three factors $(A / B / C)$ that affect the performance of a queuing system. Later, this standard was extended to three more notations, i.e. $A / B / C : X / Y / Z$. The meaning of each of these symbols is illustrated in Table 7.1. For analysis of queuing models, $A / B / C : X$ standard notation is preferred.

Figure 7.2 highlights the execution process of the queuing model that takes a series of transmission requests in the form of a queue. The pattern of the incoming queue is analyzed by the system to generate arrival (λ) and service (μ) rate. After processing requests, the queue produces relevant information, including expected number of requests in the system (L_s) and queue (L_q), along with their corresponding waiting times, i.e. W_s and W_q, respectively.

$M / M / 1 : \infty$ Queuing Model

This is one of the most basic model that assumes Poisson arrival with exponentially distributed inter-arrival epochs. The service time is also considered to be Poisson process, having a single processor to serve communication requests. Let $\lambda(x)$ and $\mu(x)$ denote the arrival and service rate to $M / M / 1 : \infty$ queuing node, with unlimited burring capacity.

$$\lambda(x) = \begin{cases} \lambda & \begin{aligned} & x \in C \\ & C = \{0,1,2,\ldots,\infty\} \end{aligned} \\ \\ 0 & otherwise \end{cases} \tag{7.15}$$

$$\mu(x) = \begin{cases} \mu & x \in C - \{0\} \\ \\ 0 & otherwise \end{cases} \tag{7.16}$$

TABLE 7.1

Queuing Model Notation and Its Interpretation

S.No.	Notation	Meaning
1.	A	Distribution of inter-arrival time to the queuing node. This describes the arrival pattern to the queue, which may follow Markovian, Erlang, degenerate, or phase-type distribution.
2.	B	Distribution of service time incurred in processing packet request at the node. It may follow distributions as specified for A. This factor is often an indication of the number of requests awaiting service at any moment of time.
3.	C	Denotes the number of processors available to serve transmission requests. In the case of wireless sensor nodes, a single processor is generally available, i.e. $C=1$
4.	X	Indicates total number of transmission requests allowed to exist in the queuing system of the node, including the ones that are currently being served. This is usually expressed as $C+X$. If this parameter is considered ∞, it implies that the queuing node can buffer unlimited transmission requests.
5.	Y	Refers to the size of packet generation source, i.e. total number of packets generated for the purpose of transmission to destined location. If this parameter is considered ∞, it implies that source is capable of producing packets infinitely.
6.	Z	Symbolizes the order or preference in which incoming transmission requests are required to be served, i.e. first-come-first-served basis, last-come-first-served basis, purely random order, or priority queue, etc.

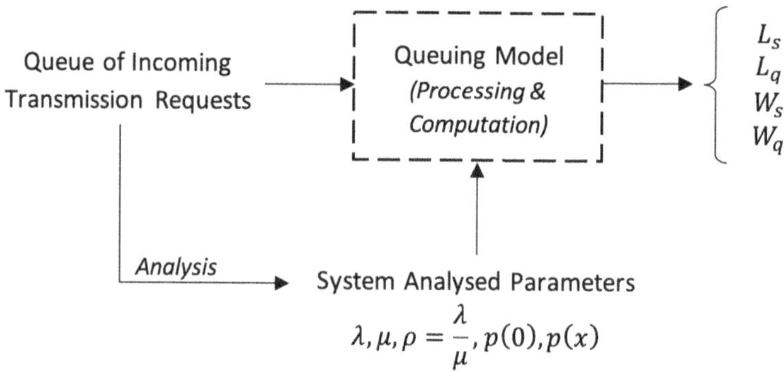

FIGURE 7.2   Queuing model and process

The queuing process progresses as depicted in Figure 7.3. In general, as a queuing system initiates, it has to evolve through a certain number of transformations and transitions. However, with progress in time and function, the system tends to attain stability. This happens because of the fact that the network commencement phase

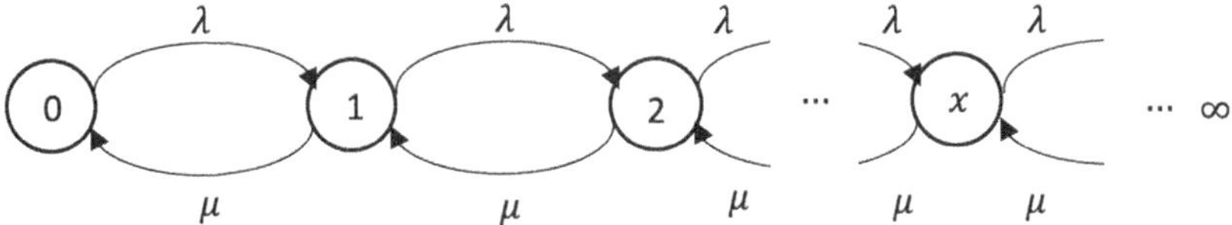

FIGURE 7.3   State transition diagram for *M/M/*1: ∞ queuing model

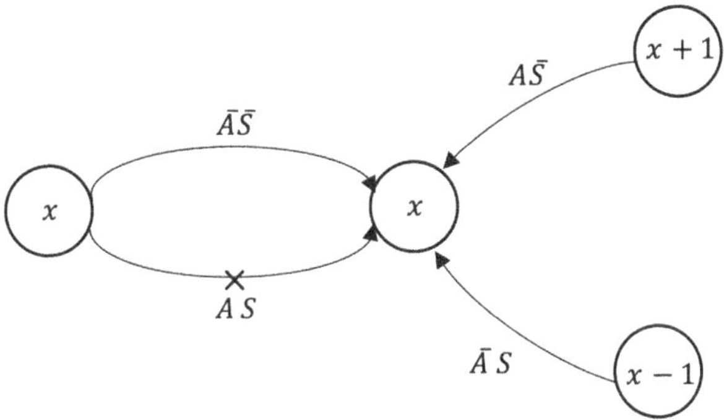

FIGURE 7.4   Steady state process for *M/M/*1: ∞ queuing model

is largely influenced by various initial conditions, for instance, initial number of requests, elapsed time, etc. This phase of transition is referred to as transient phase. Nevertheless, as sufficient time period is exceeded, the queuing system gradually becomes independent of initial conditions and enters a balanced state, better known as a steady state. In order to analyze the steady state probability of queuing systems staying in state x, Figure 7.3 illustrates a small time step " . The different ways to attain state x is analyzed with request arrival (A) or service (S). The analysis is based on the assumption that the time step is quite small, and that two events cannot take place simultaneously.

From the steady state diagram in Figure 7.4, the probability for the queuing node to remain in state x becomes

$$p_{t+\Delta}(x) = \left\{ p_t(x-1)\cdot\lambda(x-1)\Delta\cdot\left(1-\mu(x-1)\Delta\right)\right\}$$

$$+\left\{ p_t(x+1)\cdot\left(1-\lambda(x+1)\Delta\right)\cdot\mu(x+1)\Delta\right\} \qquad (7.17)$$

$$+\left\{ p_t(x)\cdot\left(1-\lambda(x)\Delta\right)\cdot\left(1-\mu(x)\Delta\right)\right\}$$

In the above equation, the term $\lambda(x-1)\Delta$ denotes the arrival rate for a small time step (Δ) w.r.t. state $(x-1)$. For simplicity, state indices are removed, i.e. $\lambda(\sim)\rightarrow\lambda$ and $\mu(\sim)\rightarrow\mu$, and higher powers of Δ are also ignored, as in the following set of equations.

$$p_{t+\Delta}(x) = \lambda\Delta p_t(x-1) + \mu\Delta p_t(x+1) + p_t(x)(1-\lambda\Delta-\mu\Delta)$$

$$\Rightarrow \frac{p_{t+\Delta}(x) - p_t(x)}{\Delta} = \lambda p_t(x-1) + \mu p_t(x+1) - p_t(x)(\lambda+\mu) \tag{7.18}$$

For a steady state case, probability does not change with respect to small changes in time step, Δ. Therefore equating $\dfrac{p_{t+\Delta}(x) - p_t(x)}{\Delta} = 0$ and adding state indices, the following is obtained:

$$\lambda(x-1)p_t(x-1) + \mu(x+1)p_t(x+1) = p_t(x)(\lambda(x)+\mu(x)) \tag{7.19}$$

Now, computing probabilities for particular values of x, we get:

$$x = 0 : \mu p_t(1) = \lambda p_t(0) \Rightarrow p_t(1) = \frac{\lambda}{\mu} p_t(0) \tag{7.20}$$

$$x = 1 : \lambda p_t(0) + \mu p_t(2) = (\lambda+\mu) p_t(1) \Rightarrow p_t(2) = \frac{\lambda^2}{\mu^2} p_t(0) \tag{7.21}$$

Now, we perform an estimation of $p_t(0)$, putting $\rho = \dfrac{\lambda}{\mu}$ and ignoring the time index, t. By probability theory, $p(0) + p(1) + p(2) + \ldots\infty = 1$. Now replacing the values of $p(x)$ in terms of $p(0)$ as computed above for different values of $x = 1, 2, \ldots$.

$$p(0) + \rho \cdot p(0) + \rho^2 \cdot p(0) + \rho^3 \cdot p(0) + \ldots\infty = 1 \tag{7.22}$$

$$\Rightarrow p(0)\left[1 + \rho + \rho^2 + \rho^3 + \ldots\infty\right] = 1 \Rightarrow p(0)\left(\frac{1}{1-\rho}\right) = 1 \Rightarrow p(0) = 1 - \rho \tag{7.23}$$

Plugging the value of $p(0)$ in a generalized case, we obtain $p(x) = \rho^x(1-\rho)$, which refers to the limiting probability (also known as steady state probability) for the queuing node to stay in state x. In other words, $p(x)$ denotes the probability that an arriving request might possibly have to wait in the system before it can be processed. This analytical study is further extended to other operating characteristics involving quantitative and temporal measures. Quantitative issues mainly focus on average communication requests residing within the system, L_s, and pending within buffer queue, L_q. Temporal measures involve the average time spent in waiting in the system, W_s, and queue, W_q. In order to analyze the mean number of requests residing in the system, L_s can be formulated as:

$$L_s = \sum_{j=0}^{\infty} j p(j) = \sum_{j=0}^{\infty} j \cdot \rho^j \cdot p(0) = p(0) \cdot \rho \sum_{j=0}^{\infty} j \cdot \rho^{j-1} \qquad (7.24)$$

$$\Rightarrow L_s = p(0) \cdot \rho \sum_{j=0}^{\infty} \frac{d}{d\rho}\left(\rho^j\right) = p(0) \cdot \rho \cdot \frac{d}{d\rho} \sum_{j=0}^{\infty}\left(\rho^j\right) = p(0) \cdot \rho \cdot \frac{d}{d\rho}\left\{1+\rho+\rho^2+\ldots\infty\right\}$$

$$\Rightarrow L_s = p(0) \cdot \rho \cdot \frac{d}{d\rho}\left(\frac{1}{1-\rho}\right) = (1-\rho) \cdot \rho \cdot \frac{1}{(1-\rho)^2} = \frac{\rho}{1-\rho} \qquad (7.25)$$

Now, mean number of packets in queue L_q is related to L_s by the following equations:

$$L_s = L_q + Expected\ no.of\ requests\ in\ service = L_q + \rho \qquad (7.26)$$

$$\Rightarrow L_q = L_s - \rho = \frac{\rho}{1-\rho} - \rho \Rightarrow L_q = \frac{\rho^2}{1-\rho} \qquad (7.27)$$

According to Little's law, average requests in the queue (or system) are computed as the product of arrival rate, λ, and expected time spent while waiting in the queue/system, i.e. $L = \lambda W$. On applying this law, following equations are obtained:

$$W_s = \frac{L_s}{\lambda} = \frac{1}{\mu-\lambda} \qquad (7.28)$$

$$W_q = \frac{L_q}{\lambda} = \frac{\lambda}{\mu(\mu-\lambda)} \qquad (7.29)$$

M / M / 2 : ∞ Queuing Model

This model should have two servers (Figure 7.5) for attending the infinite incoming requests with arrival $\lambda(x)$ and service rate $\mu(x)$ given by following equations ($x \in C$).

$$\lambda(x) = \begin{cases} \lambda & C = \{0,1,2,\ldots,\infty\} \\ \\ 0 & otherwise \end{cases} \qquad (7.30)$$

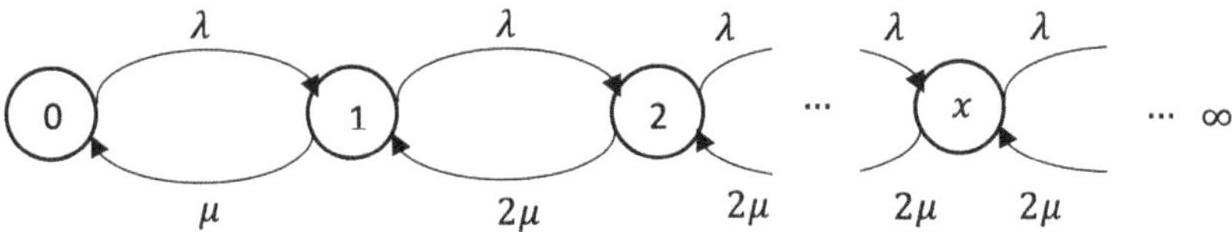

FIGURE 7.5   State diagram for *M/M/2*: ∞ queuing model

$$\mu(x) = \begin{cases} \mu x & 0 < x \leq 2 \\ 2\mu & x > 2 \\ 1 & \\ 0 & \textit{otherwise} \end{cases} \tag{7.31}$$

Similar to $M/M/1:\infty$ model derivation, the steady state distribution for $M/M/2:\infty$ computes the limiting probability, $p(x) = \left(\rho^x / 2^{x-1}\right) \cdot p(0)$ for $x \geq 2$. The probability that the queuing model is idle becomes $p(0) = (2 - \rho)/2$. Therefore, $p(x)$ becomes $\left(\dfrac{\lambda}{\mu}\right)^x \times \dfrac{1}{2^{x-1}} \times \dfrac{2\mu - \lambda}{2\mu}$.

$M/M/C:C$ Queuing Model

This generic model possesses C servers with buffer length C, i.e. no more than C requests can be accommodated in the queue [33]. Hence, as the system transits from state 0 to 1, it implies that there is presently one request in the queue, which gets served by one out of C with service rate $\mu(x)$. This implies that k servers simultaneously operating will provide a rate of $k\mu$ (Figure 7.6). The arrival and service rate to $M/M/C:C$ queuing node, having limited capacity, is:

$$\lambda(x) = \begin{cases} \lambda & C > x \geq 0 \\ 0 & \textit{otherwise} \end{cases} \tag{7.32}$$

$$\mu(x) = \begin{cases} \mu x & C \geq x > 0 \\ 0 & \textit{otherwise} \end{cases} \tag{7.33}$$

Further, the steady state probability $p(x)$ can be calculated as in above cases, which finally evaluate to:

$$p(x) = \begin{cases} \dfrac{\left(\dfrac{\lambda}{\mu}\right)^x}{x!} p(0) & 0 < x \leq C \\ 0 & \textit{otherwise} \end{cases} \tag{7.34}$$

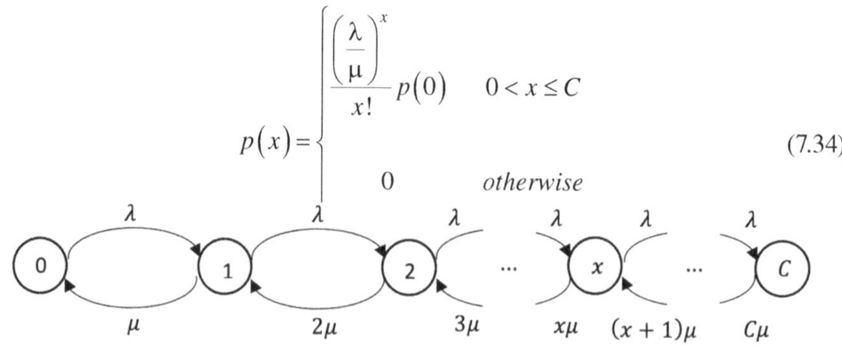

FIGURE 7.6   State diagram for $M/M/2:\infty$ queuing model

Here, $\sum_{x=0}^{C} p(x) = 1$, which helps in evaluating $p(0)$ as follows:

$$p(0)\sum_{x=0}^{C}\frac{\left(\frac{\lambda}{\mu}\right)^{x}}{x!} = 1 \Rightarrow p(0) = \left(\sum_{x=0}^{C}\frac{\left(\frac{\lambda}{\mu}\right)^{x}}{x!}\right)^{-1} \tag{7.35}$$

7.3.4 STATIC PRIORITY AND ALOHA NETWORKS

This section provides an in-depth study of performance of some real implemented protocols in wireless sensor networks that include static priority protocol (SPP), and pure and slotted Aloha networks [34].

7.3.4.1 Static Priority Protocol

In static priority scheme, the packet sending requests in lower priority queues are not attended until requests in higher priority queues are pending (Figure 7.7). This scheduling technique offers service to the incoming population of requests by managing them as classes of priorities. This scheme is practically applicable in cases that involve a single resource (or outgoing link) that needs to be accessed on a sharing basis by multiple competing requests. Initial assumptions include N priority classes, each possessing its own queue of length B_i to buffer arriving demands. Also, class 1 should be of highest priority, which gradually decreases towards class N. Each class can be modelled in the form of $M/M/1:B_i$ set of queues having arrival and departure (or service) rate as λ_i and μ_i, respectively. Owing to limited queue length, the requests exceeding buffer size are discarded. However, the highest priority queue, i.e. class 1, is exceptional and hence does not suffer from data losses, thereby offering a guaranteed service. This leads to a probability of arrival, $p_{1,arr} = \lambda_1$, and departure

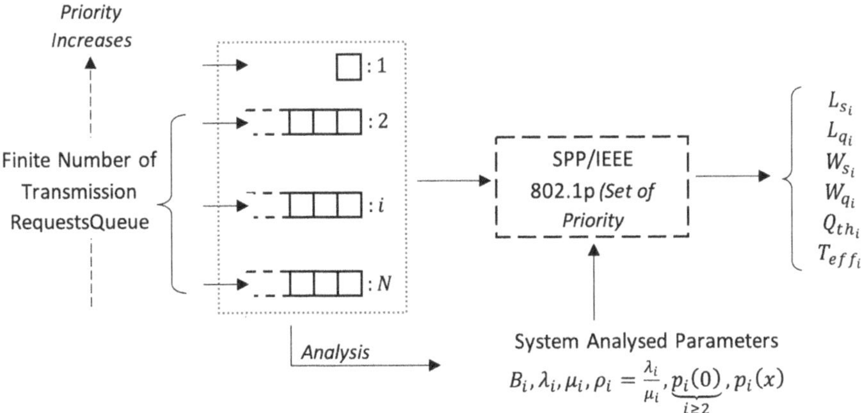

FIGURE 7.7   Static priority protocol

probability, $p_{1,dep} = \mu_1 = 1$. However, for the queue belonging to class 2, $p_{2,arr} = \lambda_2$ and $p_{2,dep} = \mu_2 = z_1 = 1 - \lambda_1$. Here, z_1 refers to the probability that class 1 is empty, i.e. no request is pending/exits.

Similarly, for class 3 we obtain $p_{3,arr} = \lambda_3$, and respective departure probability is derived as:

$$p_{3,dep} = \mu_3 = z_1 \times z_2 = \left(1 - \lambda_1\right) \times \left(\frac{1 - \rho_2}{1 - \rho_2^{B_2+1}} \right) \tag{7.36}$$

Here, z_2 is similar to the implication provided by $p_2(0) = \dfrac{1 - \rho_2}{1 - \rho_2^{B_2+1}}$ for the $M/M/1 : B_2$ queue. In general, the following can be easily summarized as $p_{i,arr} = \lambda_i$.

$$p_{i,dep} = \mu_i = z_1 \times z_2 \times \ldots \times z_{i-1} = \prod_{j=1}^{i-1} z_j \; and \; z_i = \frac{1 - \rho_i}{1 - \rho_i^{B_i+1}} \tag{7.37}$$

7.3.4.2　Pure Aloha Network

The Aloha protocol originated at the University of Hawaii to achieve the purpose of allowing wirelessly networked systems located over the wide island to communicate using wireless broadcasts. Pure Aloha offers packet transmissions with a simple technique to resolve collision. The protocol was initially executed with the assumption that transmissions can occur any time, in absence of prior carrier sensing or post acknowledgements to confirm a free channel. All the systems access common bandwidth resources without any central coordination and global time synchronization. There arises the chance of collisions whenever two or more transmissions occur simultaneously. This phenomenon of collision while accessing some shared resource is termed contention. Execution steps followed by pure Aloha is highlighted in Algorithm 7.1.

Algorithm 7.1: Execution Steps of Pure Aloha

Procedure:
Frame F arrives in queue for transmission
Transmit F
Wait for acknowledgement, F_{ack} confirming receipt of F
If F_{ack} is received, then:
Success
Exit
Else:
Wait for time, $t \in T_{random}$
Repeat step 2

When a frame is transmitted, the starting timestamp is referred to as head of the frame, and the successful reception timestamp is called tail of the frame. This suggests that a collision can occur if a frame collides with the tail or head of an ongoing transmission. Therefore, a transmission is considered successful if no two packets are allowed to be sent within a vulnerable period of $2T$ time units. Here, T refers to the time incurred in the form of transmission delay. Let N be the number of users existing in the Aloha system, and probability that a user transmits a frame is a. Contention is said to occur if a packet frame is sent at time t, while ongoing transmissions are within period $T{-}t$ and $T{+}t$. On collision, frames that become corrupt or lost are retransmitted after some random period of time.

As shown in Figure 7.8, the Aloha network can be analyzed with Markov chain by modelling in the form of three states: idle $\left(s_1\right)$, collided $\left(s_2\right)$, and transmitting $\left(s_3\right)$. The network state remains idle if none of the users are transmitting. The state gets transited to s_2 if exactly one user is sending packets. However, if two or more number of users attempt to transmit, then the state changes to s_3, i.e. collided state. While in state s_3, any arriving request would keep the system in the same state. The probability that k users demand access to the Aloha system of networked nodes during some given time step follows binomial distribution and is defined as $p_k = \binom{N}{k} a^k \left(1-a\right)^{N-k}$.

The transition matrix can be built according to the constraints followed by Aloha system, as in the following equation:

$$T_{pure} = \begin{bmatrix} p_0 & p_0 & p_0 \\ 1-p_0-p_1 & 1-p_0 & 1-p_0 \\ p_1 & 0 & 0 \end{bmatrix} \tag{7.38}$$

This study can be extended further for deriving individual state equations with the fact that state probabilities sum to 1. Therefore, solving $T_{pure} \pounds s = s_i$ becomes:

$$p_0 s_1 + p_0 s_2 + p_0 s_3 = s_1 \tag{7.39}$$

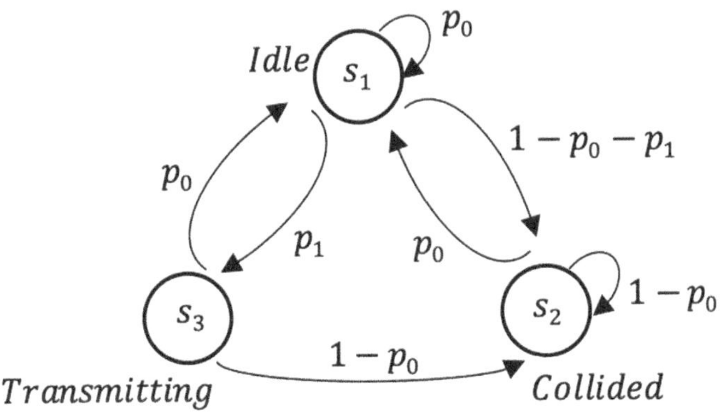

FIGURE 7.8   State transition diagram for Aloha network

$$\left(1 - p_0 - p_1\right)s_1 + \left(1 - p_0\right)s_2 + \left(1 - p_0\right)s_3 = s_2 \tag{7.40}$$

$$p_1 s_1 = s_3 \tag{7.41}$$

As we have $s_1 + s_2 + s_3 = 1$, so $s_1 = p_0$. Further, from the above equations, s_2 is evaluated to $\left[\dfrac{\left(1 - p_0\right)}{p_0} - p_1\right]p_0 = s_2 \Rightarrow s_2 = 1 - p_0 - p_0 p_1$ and $s_3 = p_0 p_1$. Also, we know that:

$$p_0 = \binom{N}{0} a^0 \left(1 - a\right)^{N-0} = \left(1 - a\right)^{N} \tag{7.42}$$

$$p_1 = \binom{N}{1} a \left(1 - a\right)^{N-1} = Na\left(1 - a\right)^{N-1} \tag{7.43}$$

Therefore, throughput for Aloha network can be given by the state equation, $s_3 = p_0 p_1$. This is because state s_3 represents the packet transmitted successfully without getting collided, corrupted, or lost, i.e. $Q_{th,pure} = s_3 = Na\left(1 - a\right)^{2N-1}$. Also, efficiency becomes

$$T_{eff,pure} = \frac{Q_{th,pure}}{A_{in,pure}} = \frac{Na\left(1 - a\right)^{2N-1}}{Na} = \left(1 - a\right)^{2N-1} \approx e^{-2aN} \tag{7.44}$$

In the above equation, the average input traffic $A_{in,pure}$ can be defined as the product of total number of possible users participating in the communication process $\left(N\right)$ simultaneously and the probability of one user transmitting, i.e. a.

7.3.4.3 Slotted Aloha Network

Slotted Aloha was designed as an upgraded model of pure Aloha protocol with certain modifications (Figure 7.9). The users in the slotted version are permitted to communicate at the beginning of time step T, unlike its pure version in which a user was given the independence of scheduling its transmission at random times. However, if a user fails to avail of the opportunity of transmitting at the start of a particular time frame, it waits for the start of the next time frame. Owing to this constraint, slotted Aloha can support transmission of two packets in time $2T$. Therefore, the vulnerable time improves for the slotted protocol as T.

According to the transition diagram in Figure 7.9 , the following transition diagram can be built.

$$T_{slotted} = \begin{bmatrix} p_0 & p_0 & p_0 \\ 1 - p_0 - p_1 & 1 - p_0 - p_1 & 1 - p_0 - p_1 \\ p_1 & p_1 & p_1 \end{bmatrix} \tag{7.45}$$

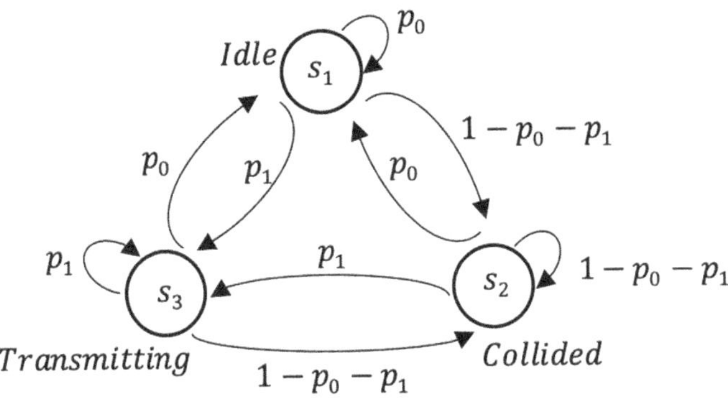

FIGURE 7.9   State transition diagram for pure Aloha network

Resultantly, state equations become $s_1 = p_0$, $s_2 = 1 - p_0 - p_1$, and $s_3 = p_1$. Also throughput can be given by the state $s_3 = Q_{th,slotted} = Na(1-a)^{N-1} \approx Nae^{-Na}$. The efficiency of slotted Aloha protocol is given by

$$T_{eff,pure} = \frac{Q_{th,pure}}{A_{in,pure}} = \frac{Nae^{-Na}}{Na} \approx e^{-aN} \tag{7.46}$$

Conclusively, though slotted Aloha have identical probability, p_0 for remaining in s_1 state, as compared to its pure version but enjoys a privilege of having lesser probability, $1 - p_0 - p_1$, for being in s_2, i.e. colliding state.

7.3.5   NETWORK SIMULATION

In general, simulation is a key component in research that offers modelling of some real system, protocol, or process. Such models can be communication standards, protocols, or architectures that offer new technologies, upgraded modules, and security solutions. This helps in the study of behaviour of proposed research under time and resource constraints.

Network simulation is a significant approach to evaluate networking concepts and methodologies in the context of wired as well as wireless networks. There are several network simulators available that mainly include NS-2, NS-3, and OMNet++ [35, 36]. These simulation tools are equipped with core-level programming platform, user-friendly interfaces, and rich animation for communication visualization [37]. Such features enable the network developers to innovate novel ideas, experiment with execution process, and monitor outcomes. Moreover, frequent experimentation with real world systems can be expensive and incur risk, which further necessitates simulation of new approaches before on-field implementation.

The simulators allow modelling arbitrary, non-existent networking platforms and evaluate their performance by specifying parameters relevant to wired/wireless nodes and communication channels. The simulators impart freedom to the developers to design different networking topologies and scenarios to examine the impact of the architectural layout of nodes on the performance of proposed models. Programmers can then investigate the wide characteristics of networks by varying a set of simulation parameters and associated environmental variables.

In addition to immense networking abilities, simulation requires some general capabilities, including a debugging facility, code modularity, and reproducibility without depending on underlying software and hardware configurations. Some key issues involved in the correctness of computer simulation include procurement of valid datasets, choice of parameters, application of mathematical approximations, and suitable use of randomization to ensure validity and reliability of simulation results.

7.3.5.1 NS-2

NS-2 is one of the most basic platforms that offers simulation of wired, wireless, and hybrid networking platform. It is available as open-source, making it widely used by network programmers along with its support for core programming with C++ and oTcl scripts. NS-2 offers sequential simulation with backend protocol development in C++ language, which assists in modelling characteristics and predicting the behaviour of simulation nodes. Furthermore, oTcl scripts are used to monitor the simulation environment and control network parameters, topological specifications, etc. for instance the number of nodes, simulation time, type of buffer queue, queue length, and channel bandwidth. As a major drawback, this simulator requires frequent recompilation of all backend C++ programs to reflect updates of newly added protocols. This activity is considered time-consuming and decelerates the research cycle.

7.3.5.2 NS-3

Similar to its preceding version, NS-3 is also based on C++ programming for protocol design and development. However, NS-3 relies on Python for frontend programming, which is required for managing the simulation outcome, unlike NS-2 which used oTcl scripts and confronted several problems due to the combination of C++ coding with those scripts. The simulations carried out with NS-3 require mainly C++ with optional usage of Python. Furthermore, the simulator incorporates architectural models from GTNetS that originally contained an extended support for scalability issues. The features and performance issues were extended to match compatibility with programming needs.

7.3.5.3 OMNeT++

Unlike the above simulators, OMNeT++ is referred to as a general purpose discrete event-based simulation (DES) platform that includes a huge collection of communication protocols and internet models. It also extends a comprehensive support for the mobility framework and Castalia, which further assists in the especially wireless and ad hoc network simulation at even a larger scale. OMNeT++ consists of a rich modularity that enables ease in analyzing the behaviour of individual modules along with better debugging. This further enriches the development of protocols that are especially designed to serve cross-layer and cross-platform goals. It provides the capability to link small modules exhibiting a distinct feature of a protocol, thereby forming compound modules.

Moreover, any simulation programmed in OMNeT++ can itself be implemented in the form of a compound module which might further be incorporated with other simple or compound modules. The design of simple modules in OMNet++ is performed with the help of rich object-oriented C++ libraries, while composition of individual simple modules into compound ones is achieved with NED (Network Description language). During compilation, NED is patently rendered into C++ programs for efficient simulation. Also, OMNeT++ supports dynamic declaration of certain parameters, such as number of sensors or host/sink nodes in the run-time environment.

7.3.5.4   JiST

A Java-based prevalent simulator in the field of networking is JiST (Java in Simulation Time). JiST is normally used in concurrence with another simulator called SWANS, which was explicitly built to promote ad hoc network simulation with mobile nodes. The prevalence of JiST is mainly due to the efficiency in simulation time in terms of rapidity as well as accuracy. The simulator operates with simplest network entities called nodes. The processes coded within an entity are implemented in form of arbitrary Java programs. The interactions among entities act as benchmarks for synchronizing events to facilitate faster execution of codes simultaneously over individual entities, thereby leading to better performance. However, the potential improvement in simulation speed is achieved at the expense of higher computational cost and memory usage, which eventually poses a major issue of concern in judging its overall performance.

7.3.5.5   SimPy

SimPy is another simulator that is widely known for its process-oriented approach along with support for discrete-event simulation. The programming interface of the application is coded in Python, which considers each simulation entity as an individual process. These processes can be executed simultaneously and also are capable of simulating message exchanges in the form of object passing between processes. Additionally, SimPy is capable of synchronizing the processes during simulation for the purpose of efficiently monitoring the simulation time stamped data.

7.4   SELF-SIMILARITY PHENOMENA IN SENSOR NETWORKS

Self-similarity is a phenomena where patterns at various time scales show similarities, and it is seen in particular types of data or processes. Self-similarity can be seen in the traffic patterns produced over time by the sensor nodes in sensor networks [38]. When data traffic displays self-similarity, it indicates that the pattern of traffic at a smaller time scale (for example, short time intervals) reflects the pattern of traffic at a larger time scale (for example, longer time intervals).

Self-similarity phenomena in sensor networks can have several implications and effects that are highlighted as follows.

a. Traffic burstiness: Self-similar traffic frequently displays bursty behaviour, in which bursts of intense activity are followed by spans of inactivity. The allocation of resources and capacity-planning for the network may be impacted by the sporadic nature of the traffic.

b. Long-range dependence: Self-similar traffic exhibits long-range dependence, indicating that there may be a considerable correlation between traffic at remote time intervals. Unpredictable traffic peaks and variations may result from long-range dependency.

c. Network performance: Throughput, latency, and packet loss are few of the performance indicators of a network that self-similar traffic can significantly affect. Self-similar traffic may not behave as expected in traditional network models that presume traffic to be Poisson or Markovian.

d. Buffer overflow: The buffer widths in routers or switches must be created to effectively tolerate bursty traffic patterns when there is self-similar traffic. Otherwise, packet loss and buffer overflow may happen.

e. Traffic engineering: Self-similarity in sensor network traffic necessitates careful capacity planning and traffic engineering to guarantee effective network resource utilization and prevent bottlenecks. The bottleneck nodes are often considered primary causes for congestion and delayed transmission.

f. Service quality management: Self-similar traffic can make it difficult to manage quality of service because of the bursty nature, which may cause variations in service performance.

g. Traffic modelling: For the purpose of creating effective communication protocols and forecasting network behaviour under varied circumstances, accurate modelling of self-similar traffic is crucial.

Understanding and accounting for self-similarity phenomena are crucial for designing and managing sensor networks effectively. Traditional networking techniques that assume traffic to be Poisson or constant bit rate may not accurately represent the actual traffic behaviour in self-similar scenarios. Specialized traffic models, such as fractional Brownian motion and long-range dependent (LRD) models, are used to better capture the self-similarity characteristics in sensor network traffic. By incorporating self-similarity into network modelling and resource allocation, engineers and researchers can better optimize the performance and efficiency of sensor networks and ensure their robustness in handling real-world traffic patterns.

REFERENCES

1. Padmavathy, M. Chitra, and M. Chitra. "Performance Evaluation of Energy Efficient Modulation Scheme and Hop Distance Estimation for WSN." *International Journal of Communication Networks and Information Security* 2, no. 1 (2010): 44–9.

2. Zordan, Davide, Borja Martinez, Ignasi Vilajosana, and Michele Rossi. "On the Performance of Lossy Compression Schemes for Energy Constrained Sensor Networking." *ACM Transactions on Sensor Networks (TOSN)* 11, no. 1 (2014): 1–34.
3. Min, Hla Yin, and Win Zaw. "Performance Evaluation of Energy Efficient Cluster-Based Routing Protocol in Wireless Sensor Networks." *International Journal of Computer Science Engineering IJCSE* 3, no. 2 (2014): 71–6.
4. Senouci, Mustapha Reda, Abdelhamid Mellouk, Hadj Senouci, and Amar Aissani. "Performance Evaluation of Network Lifetime Spatial-Temporal Distribution for WSN Routing Protocols." *Journal of Network and Computer Applications* 35, no. 4 (2012): 1317–28.
5. Kabara, Joseph, and Maria Calle. "MAC Protocols Used by Wireless Sensor Networks and a General Method of Performance Evaluation." *International Journal of Distributed Sensor Networks* 8, no. 1 (2012): 834784.
6. Rong, Bo, Yi Qian, Mahamat H. Guiagoussou, and Michel Kadoch. "Improving Delay and Jitter Performance in Wireless Mesh Networks for Mobile IPTV Services." *IEEE Transactions on Broadcasting* 55, no. 3 (2009): 642–51.
7. Phan, Linh-An, Taejoon Kim, Taehong Kim, JaeSeang Lee, and Jae-Hyun Ham. "Performance Analysis of Time Synchronization Protocols in Wireless Sensor Networks." *Sensors* 19, no. 13 (2019): 3020.
8. Devi, V. Seedha, T. Ravi, and S. Baghavathi Priya. "Cluster Based Data Aggregation Scheme for Latency and Packet Loss Reduction in WSN." *Computer Communications* 149 (2020): 36–43.
9. Munir, Sirajum, Shan Lin, Enamul Hoque, SM Shahriar Nirjon, John A. Stankovic, and Kamin Whitehouse. "Addressing Burstiness for Reliable Communication and Latency Bound Generation in Wireless Sensor Networks." In Proceedings of the 9th ACM/ IEEE International Conference on Information Processing in Sensor Networks, pp. 303–14, 2010.
10. Zhao, Jerry, and Ramesh Govindan. "Understanding Packet Delivery Performance in Dense Wireless Sensor Networks." In Proceedings of the 1st International Conference on Embedded Networked Sensor Systems, pp. 1–13, 2003.
11. Bhadra, Dhwani R., Charmi A. Joshi, Priya R. Soni, Nikita P. Vyas, and Rutvij H. Jhaveri. "Packet Loss Probability in Wireless Networks: A Survey." In *2015 International Conference on Communications and Signal Processing (ICCSP)*, pp. 1348–54, IEEE, 2015.
12. Saini, Preeti, Rishi Pal Singh, and Adwitiya Sinha. "Path Loss Analysis of RF Waves for Underwater Wireless Sensor Networks." In 2017 International Conference on Computing and Communication Technologies for Smart Nation (IC3TSN), pp. 104–8, IEEE, 2017.
13. Liu, Wang, Kejie Lu, Jianping Wang, Guoliang Xing, and Liusheng Huang. "Performance Analysis of Wireless Sensor Networks with Mobile Sinks." *IEEE Transactions on Vehicular Technology* 61, no. 6 (2012): 2777–88.
14. Saini, Preeti, Rishi Pal Singh, and Adwitiya Sinha. "Path Loss Analysis of RF Waves for Underwater Wireless Sensor Networks." In 2017 International Conference on Computing and Communication Technologies for Smart Nation (IC3TSN), pp. 104–8, IEEE, 2017.
15. Taddia, C., and G. Mazzini. "On the Retransmission Methods in Wireless Sensor Networks." In IEEE 60th Vehicular Technology Conference, 2004, VTC2004-Fall 2004, vol. 6, pp. 4573–7, IEEE, 2004.

16. Suma, S., and Bharati Harsoor. "Dynamic Shortest Path Routing Algorithm to Reduce Retransmission and Congestion Avoidance for Mobile Nodes in Wireless Sensor Network." In IoT Based Control Networks and Intelligent Systems: Proceedings of 3rd ICICNIS 2022, pp. 649–59. Singapore: Springer Nature Singapore, 2022.

17. Manuel, Ebin M., Vinod Pankajakshan, and Manil T. Mohan. "Data Aggregation in Low-Power Wireless Sensor Networks with Discrete Transmission Ranges: Sensor Signal Aggregation over Graph." *IEEE Sensors Journal* 22, no. 21 (2022): 21135–44.

18. Sinha, Adwitiya, and D. K. Lobiyal. "Prediction Models for Energy Efficient Data Aggregation in Wireless Sensor Network." *Wireless Personal Communications* 84 (2015): 1325–43.

19. Sinha, Adwitiya, and D. K. Lobiyal. "Probabilistic Data Aggregation in Information-Based Clustered Sensor Network." *Wireless Personal Communications* 77 (2014): 1287–310.

20. Manuel, Ebin M., Vinod Pankajakshan, and Manil T. Mohan. "Energy-Efficient Data Aggregation in Low-Power Wireless Networks with Sensors of Discrete Transmission Ranges: A Mathematical Framework for Network Design." *IEEE Transactions on Network Science and Engineering* (2023). https://doi.org/10.1109/tnse.2023.3274693.

21. Sinha, Adwitiya, and D. K. Lobiyal. "A Multi-Level Strategy for Energy Efficient Data Aggregation in Wireless Sensor Networks." *Wireless Personal Communications* 72 (2013): 1513–31.

22. Correia, Felipe, Marcelo Alencar, and Karcius Assis. "Stochastic Modeling and Analysis of the Energy Consumption of Wireless Sensor Networks." *IEEE Latin America Transactions* 21, no. 3 (2023): 434–40.

23. Li Jianpo, Qing Han, and Wenting Wang. "Characteristics Analysis and Suppression Strategy of Energy Hole in Wireless Sensor Networks." *Ad Hoc Networks* 135 (2022): 102938.

24. Vasanthi, G., and N. Prabakaran. "An Improved Approach for Energy Consumption Minimizing in WSN Using Harris Hawks Optimization." *Journal of Intelligent & Fuzzy Systems* 43, no. 4 (2022): 4445–56.

25. Carsancakli, Muhammed Fatih, Md Abdullah Al Imran, Huseyin Ugur Yildiz, Ali Kara, and Bulent Tavli. "Reliability of linear WSNs: A Complementary Overview and Analysis of Impact of Cascaded Failures on Network Lifetime." *Ad Hoc Networks* 131 (2022): 102839.

26. Jain, Khushboo, Anoop Kumar, and Akansha Singh. "Data Transmission Reduction Techniques for Improving Network Lifetime in Wireless Sensor Networks: An up-to-Date Survey from 2017 to 2022." *Transactions on Emerging Telecommunications Technologies* 34, no. 1 (2023): e4674.

27. Jaiswal, Priyanka, and Adwitiya Sinha. "Stable Geographic Forwarding with Link Lifetime Prediction in Mobile Adhoc Networks for Battlefield Environment." *Human-Centric Computing and Information Sciences* 6 (2016): 1–18.

28. Jeong, Wootae, and Shimon Y. Nof. "Performance Evaluation of Wireless Sensor Network Protocols for Industrial Applications." *Journal of Intelligent Manufacturing* 19 (2008): 335–45.

29. Le Boudec, Jean-Yves. *Performance Evaluation of Computer and Communication Systems.* Lausanne: Epfl Press, 2010.

30. Raposo, Duarte, André Rodrigues, Jorge Sá Silva, and Fernando Boavida. "A Taxonomy of Faults for Wireless Sensor Networks." *Journal of Network and Systems Management* 25 (2017): 591–611.

31. Ram, Mahendra, Sushil Kumar, Vinod Kumar, Ajay Sikandar, and Rupak Kharel. "Enabling Green Wireless Sensor Networks: Energy Efficient T-MAC Using Markov Chain Based Optimization." *Electronics* 8, no. 5 (2019): 534.

32. Jiang, Fuu-Cheng, Der-Chen Huang, Chao-Tung Yang, and Fang-Yi Leu. "Lifetime Elongation for Wireless Sensor Network Using Queue-Based Approaches." *The Journal of Supercomputing* 59 (2012): 1312–35.
33. Kang, Zhi Hu, Geng Sheng Zheng, Qiang Gao, and Ao Cheng Huang. "M/M/1/n-Based MAC Protocol in Wireless Sensor Networks with Adaptive Duty-Cycle." *Advanced Materials Research* 926 (2014): 2494–8.
34. Choi, Hyun-Ho, and Wonjae Shin. "Slotted ALOHA for Wireless Powered Communication Networks." *IEEE Access* 6 (2018): 53342–55.
35. Khot, Dipti D., Amol Patole, and J. S. Awati. "Comparison Study of Ns-2 and Exata Softwares For Wsn." *International Journal of Innovations in Engineering Research and Technology*: 1–4. https://doi.org/10.22214/ijraset.2019.5594.
36. Monir, Md Fahad, and Tahmid Alavi Ishmam. "Exploiting Link Diversity in IEEE 802.11 WLAN using OMNeT++." In 2022 IEEE 10th Region 10 Humanitarian Technology Conference (R10-HTC), pp. 355–9, IEEE, 2022.
37. Qadar, Rabia, Waleed Bin Qaim, Bo Tan, and Jari Nurmi. "Underwater Optical Communication Module: An Extension to the ns-3 Network Simulator." In 2022 IEEE 96th Vehicular Technology Conference (VTC2022-Fall), pp. 1–5, IEEE, 2022.
38. Chen, Dan, Housheng Su, and Zhigang Zeng. "Geometric Renormalization Reveals the Self-Similarity of Weighted Networks." *IEEE Transactions on Computational Social Systems* 10, no. 2 (2022): 426–34.

Index